Whitworth College Library

0 0068 00012628 1

SO-CAE-795

DISCARD

Perspectives on Research
on

Effective Mathematics Teaching

Volume 1

Editors

Douglas A. Grouws
University of Missouri—Columbia

Thomas J. Cooney
University of Georgia

Technical Editor

Douglas Jones
University of Georgia

LAWRENCE ERLBAUM ASSOCIATES

NATIONAL COUNCIL OF
TEACHERS OF MATHEMATICS

Copyright © 1988 by
THE NATIONAL COUNCIL OF TEACHERS OF MATHEMATICS, INC.
1906 Association Drive, Reston, Virginia 22091
All rights reserved

ISBN: 0-87353-254-6 (Vol. 1, paper)
ISBN: 0-8058-0326-2 (Vol. 1, cloth)
ISBN: 0-87353-256-2 (5-vol. set, paper)

Second printing 1989
Third printing 1989

This material is based upon work supported by the National Science Foundation under Grant No. MDR-8550614. The Government has certain rights in this material. Any opinions, findings, and conclusions or recommendations expressed in this material are those of the author(s) and do not necessarily reflect the views of the National Science Foundation.

The publications of the National Council of Teachers of Mathematics present a variety of viewpoints. The views expressed or implied in this publication, unless otherwise noted, should not be interpreted as official positions of the Council.

Printed in the United States of America

CONTENTS

Acknowledgements

The papers and conference proceedings recorded in this monograph represent the efforts of many individuals, and we appreciate the contribution each has made. Of course, we are especially grateful to the authors and to the conference participants, whose names appear in the last section of this volume. The following doctoral students made a significant contribution to the Research on Effective Mathematics Teaching Conference as session recorders: Terry Crites, Barbara Dougherty, Doug Jones, Joanne Nelson, and Debra Perkowski. A special thank you is also due to graduate students Leroy Sharp and Somsong Donkaewbua for their assistance with many of the conference arrangements. We are grateful for the assistance provided by our line editor, Sandra Ward. Stephanie Bales provided invaluable assistance with the typing. Finally, we offer acknowledgement and appreciation for the administrative and financial support of Dr. W. R. Miller, Dean of the College of Education at the University of Missouri, and Dr. Wayne Dumas, Chairman of the Department of Curriculum and Instruction at the University of Missouri. The entire project would not have been possible without the financial support of the National Science Foundation.

Series Foreword

We clearly know more today about teaching and learning mathematics than we did twenty years ago, and we are beginning to see the effects of this new knowledge at the classroom level. This is possible in part because of the financial support that has become available to researchers. If theory building and knowledge acquisition are to have a basis broad enough to inform policy and influence educational practice, such support is essential. Although funding levels remain low in comparison to existing needs, there are several research projects either completed or in progress that could not have been undertaken without support.

In particular, we can point to several significant sets of studies based on emerging theoretical frameworks. For example, young children's early number learning and older children's understanding of rational numbers have been the subject of several recent research programs. Most of us who do research would agree that our work is more likely to be profitable when it results from an accumulation of knowledge acquired through projects undertaken within a coherent framework rather than through single, isolated studies. To establish such a framework, researchers must be provided with the opportunity to exchange and refine their ideas and viewpoints. Conferences held in Georgia and Wisconsin during the seventies serve as examples of the role such meetings can play in providing a vehicle for increased communication, synthesis, summary, and cross-disciplinary fertilization among researchers working within a specialized area of mathematical learning.

Over the past few years, the members of the Research Advisory Committee of the National Council of Teachers of Mathematics (NCTM) have observed specializations emerge that could benefit from collaborative efforts. We therefore proposed to the National Science Foundation that funding be provided for the purpose of establishing research agendas in several areas where conceptual and methodological consensus seemed possible. We believed that such a project was needed at this time for two reasons: first, to direct research efforts toward important questions, and second, to encourage the development of support mechanisms essential to collaborative chains of inquiry. Four such specialized areas were selected for this project: the teaching and assessing of problem solving, the teaching and learning of algebra, effective mathematics teaching, and the learning of number concepts by children in the middle grades.

The plan for the project included a working group conference in each of the four areas, with monographs of conference proceedings to be published by the National Council of Teachers of Mathematics. An overview monograph, written by advisory board members, was also planned. The advisory board consisted of F. Joe Crosswhite, James G. Greeno, Jeremy Kilpatrick, Douglas B. McLeod, Thomas A. Romberg, George Springer, James W. Stigler, and Jane O. Swafford, while I served as project director. For each of the four selected areas, two researchers were named to serve as conference co-directors and as co-editors of the monograph of conference proceedings. These pairs were Edward A. Silver and Randall I. Charles for teaching and assessing problem solving; Sigrid Wagner and Carolyn Kieran for learning and teaching algebra; Douglas A. Grouws and Thomas J. Cooney for effective mathematics teaching; and Merlyn J. Behr and James Hiebert for number learning in the middle grades.

The project began in May of 1986 with a planning meeting of advisory board members and conference co-directors. Issues to be addressed and possible paper topics for each conference were first identified by the group. Tentative lists of invitees for each conference were drawn up to include researchers from mathematics education and relevant fields of psychology and social science, as well as mathematicians and practitioners. The names of promising young researchers were included along with names of established researchers. An international perspective was considered important, and so the list also included names of scholars from abroad. The concept of working group conferences funded for 25 people precluded expanding the conference to all interested persons. We therefore decided to invite people to attend only one conference, thus maximizing the total number of persons involved in the project. The final participant lists follow this report.

The first of the four working conferences, on teaching and assessing mathematical problem solving, was held in January, 1987, in San Diego. Several approaches to teaching problem solving were advanced and discussed: teaching *as* problem solving, the teacher as coach versus teacher as manager, modeling master teachers, viewing students as apprentices, the use of macro-contexts to facilitate mathematical thinking, consideration of mathematics as an ill-structured discipline. Discussion of assessment issues focused on process rather than outcome, and questions of evaluating processess and schema structures were explored. Considerable attention was given throughout the conference to associated problems of teacher preparation.

The conference on effective mathematics teaching was held in March in Columbia, Missouri. The first paper delivered there, on teaching for higher-order thinking in mathematics, tied this conference to the first one. A paper on the functioning of educational paradigms set a stage for much of the discussion at the conference. The question of what makes a good mathematics teacher was explored in discussions ensuing from presentations on expert-novice studies, on cross-cultural studies, and on teacher professionalism. Concern for the content in mathematics classes and the manner in which this is determined also received attention. The lack of funding necessary for observational research was strongly noted.

The conference on teaching and learning algebra, also in March, was held in Athens, Georgia. The papers and discussions focused on four major themes: what is algebra and what should it become, in light of continuing technological advances; what has research told us about the teaching and learning of algebra; what is algebraic thinking and how does it relate to general mathematical thinking; and what is the role of representations in the learning of algebra. Research questions from the perspectives of content, of learning, of instruction, and of representations were formulated.

The final conference, on number concepts in the middle grades, was held in May in DeKalb, Illinois. A theme permeating many of the papers at this conference was that number concepts related to topics taught at this level are qualitatively different from those in lower grades, both in terms of the number systems studied and the operations involved. The papers explored the new conceptions and complexities that students encountered, and examined the effects of conventional and experimental instructional programs. It was acknowledged that the differences between early and later number concepts need to be recognized and more adequately understood before instructional programs can be developed to enhance number learning in the middle grades.

The brevity of these descriptions does not do justice to the diversity and richness of the papers and discussions. Each of the conferences was indeed a working conference. Participants addressed difficult questions and discussions were lively and intense. Consensus was elusive, as might be expected with a group of people with such diverse backgrounds. Even so, there was agreement on many fundamental issues, and individual researchers, representing different disciplines and viewpoints, were able to reach new understandings.

There are five monographs being published as a result of this project, all under the general title *Research Agenda in Mathematics Education*. Four of the monographs contain conference proceedings and are subtitled as follows: *The Teaching and Assessing of Mathematical Problem Solving, Perspectives on Research on Effective Mathematics Teaching, Research Issues in the Learning and Teaching of Algebra,* and *Number Concepts and Operations in the Middle Grades.* The proceedings include revised conference papers, some discussion, response, and summary papers by other conference participants, and chapters by the co-editors. They will be of particular interest to researchers interested in the learning and teaching of mathematics. The fifth monograph, *Setting a Research Agenda,* is intended for a wider audience, including policy makers, mathematics supervisors, and teachers. In this monograph the project advisory board discusses the past and present state of research in mathematics education and cognitive science, the relation of reform movements to research efforts, the role of this project in guiding future research in mathematics education, major issues addressed at the conferences and other issues still needing to be addressed, and resources needed to facilitate research. We are fortunate to have Lawrence Erlbaum Associates, Inc., join with the National Council of Teachers of Mathematics in publishing this series of monographs.

This project was funded by the National Science Foundation under Grant No. MDR-8550614. The advisory board and monograph editors join me in an expression of gratitude to Raymond I. Hannapel of the National Science Foundation for his continued support throughout the term of this project. We also wish to thank James D. Gates and Charles R. Hucka of the National Council of Teachers of Mathematics for their assistance with the publication of these monographs, and Julia Hough of Lawrence Erlbaum Associates for her work in facilitating joint publication between NCTM and Erlbaum. Finally, we wish to acknowledge the assistance of administrators at San Diego State University, University of Missouri, University of Georgia, and Northern Illinois University, and thank them for the many amenities they provided to conference participants.

I personally want to express my appreciation to the members of the advisory board for all the assistance they have given me during these past two years. My largest debt of gratitude is owed to the conference co-directors and co-editors for their work in directing four outstanding conferences and for providing all of us with sets of proceedings that will guide research efforts in the years to come.

Judith T. Sowder
San Diego State University

Introduction

As the era of reform in education continues, it becomes especially important that we develop a conceptually rich understanding of what effective mathematics teaching is and how to foster it. A major part of this effort must center on the teaching process itself and take into account many inextricably linked ideas, including the way in which students learn mathematics, the nature of the mathematics to be taught, underlying social conditions, and the many constraints placed directly or indirectly on classroom teachers. No longer can we uncritically rely on the folklore of the classroom; rather, we must sort out fact from myth and begin to build conceptual networks that will help us understand and improve the complex process of teaching.

The papers published in this monograph call attention to many of the important and relevant issues that must be addressed in building a framework for effective mathematics teaching. A prescriptive agenda for future research is not stated, perhaps to the dismay of some readers. To suggest a single intellectual path for developing the necessary theoretical base would be difficult, if not impossible, and probably would be counterproductive. A forward move in research on effective mathematics teaching demands investigations from a variety of perspectives.

It is our sincere hope that in your reading of the papers in this volume you will reflect on many facets of mathematics teaching and that many of you will be stimulated to conduct some research, formally or informally, alone or in collaboration with a colleague, on some aspect of mathematics instruction. The research enterprise is the responsibility of all who have an interest in effective mathematics teaching.

Teaching for Higher-Order Thinking in Mathematics: The Challenge for the Next Decade

Penelope L. Peterson

University of Wisconsin—Madison

The 1980s may go down in history of education textbooks as the decade for reports that decried the quality of public education in our country. Beginning with *A Nation At Risk*, (National Commission on Excellence in Education, 1983), the reports indicated that elementary and secondary schools in our country are not succeeding in their job of educating students. Most recently, a report by the U.S. Office of Education (1987) comparing U.S. schooling with Japanese schooling stated that improving the quality of schooling in the United States is necessary for our economic survival. Taken together, the reports highlighted the fact that elementary and secondary students are not achieving as highly as they might on basic skills and, in particular, that students are not achieving well on academic tasks that require higher-level thinking, such as problem solving in mathematics.

Because learning problems and achievement deficits that appear in the elementary school years will be magnified later in the secondary school years, the improvement of elementary teaching and schooling is a clear priority (see, for example, Bennett, 1986). This may be particularly true for higher-order thinking skills in mathematics because it is these higher-order thinking skills that enable an individual to learn more mathematics, to use mathematics in other disciplines, and to solve mathematical problems throughout life (Fennema & Peterson, 1985). Thus, these data suggest the need for an increased instructional focus on teaching higher-level skills in mathematics to all students. Such an increased focus might be particularly important for lower-achieving students, who have more difficulty than their peers in learning these higher-order skills on their own.

Overview of This Paper

Writers of reports in the 1980s have presented a clear challenge to educators for the next decade: to improve student learning and achievement, particularly student learning and achievement of higher-order thinking skills. In this chapter we focus on teaching higher-order thinking in mathematics in elementary school. We focus on mathematics because of the importance of mathematics to an individual's performance in everyday life and to an individual's economic and financial success and because recent reports have focused on mathematics learning and achievement. We begin by describing elementary mathematics classrooms today. We then discuss three dimensions of teaching and classroom learning that seem to be importantly related to students' learning of higher-order skills in mathematics in elementary school. We show how contemporary teaching practices and classroom processes may not be facilitating the development of higher-order thinking skills in mathematics. Then we discuss findings from recent research that suggest what teaching strategies and classroom processes might promote students' higher-order thinking in mathematics. We conclude with some specific suggestions for educational researchers and practitioners in the next decade for how they might address the challenge of improving students' higher-order thinking skills in mathematics.

Elementary Mathematics Classrooms Today

Romberg and Carpenter (1986) argued that mathematics classrooms today are much like those that existed when we were in school. The picture of the classroom is one of extensive teacher-directed explaining and questioning in the context of whole-group instruction followed by students working on paper-and-pencil assignments at their seats. For example, in a study of 36 fourth-grade mathematics classrooms in Wisconsin, Peterson and Fennema (1985) found that 43% of the class time in mathematics was spent in whole-group instruction, and 47% of the time was spent with students doing seatwork. Similarly, in his study of 1,000 classrooms Goodlad (1984) found that the following pattern consistently characterized teaching and classrooms regardless of the grade level or subject matter: (a) a predominance of whole-group instruction; (b) each student working and achieving alone within a group setting; (c) the teacher functioning as the central figure in determining activities and conducting instruction; (d) a predominance of frontal teaching and monitoring of students' seatwork by the teacher; and (e) students rarely actively engaged in learning directly from one another or in initiating processes of interaction with teachers.

Researchers have also documented that most of the time spent in elementary mathematics classrooms is focused on the teaching and learning of lower-level skills and concepts in mathematics rather than on higher-order thinking (e.g., Porter, Floden, Freeman, Schmidt, & Schwille, 1987). For example, Peterson and Fennema (1985) found that during fourth-grade mathematics classes, students spent only 15% of their time engaged in learning higher-level mathematics content; 62% of their time students were engaged in learning lower-level mathematics content, and 13% of the time students were not engaged in learning mathematics. In categorizing the content of mathematics teaching and classroom activities, Peterson and Fennema used the National Assessment of Educational Progress (NAEP) (1979) categories of knowledge and skill to define low-level mathematics learning and the NAEP categories of understanding and application to define high-level learning. Figure 1 shows examples of low-level and high-level mathematics content. In the discussion that follows, we use these operational definitions of low-level and high-level learning in mathematics.

In sum, the picture of elementary mathematics classrooms today is one

Effective Mathematics Teaching

Low-Level Problems   High-Level Problems

Computational
Problems

1.

WHAT IS THE LENGTH OF THIS PENCIL TO THE NEAREST INCH?

ANSWER_____

2.

HOW LONG IS THIS LINE SEGMENT?

A. 1 INCH
B. 2 INCHES
C. 3 INCHES
D. 4 INCHES
E. 5 INCHES
F. I DON'T KNOW.

Computational
Problems

3.   18 x 0 x 29 =

Answer_____

4. WRITE A NUMERAL FOR THE FOLLOWING:

3 x 100 + 5 x 10 + 9 x 1

Answer_____

Conceptual
Problems

5.   STEVE, GRACE, TONY, AND IRENE BELONG
TO THE SAME BOOK CLUB. RICK, TONY,
STEVE, AND JOHN BELONG TO THE FRENCH
CLUB. WHICH OF THESE CHILDREN ARE
IN THE INTERSECTION OF THE TWO
CLUBS?

Answer_____

6. TWO TEAM CAPTAINS TAKE TURNS CHOOSING PLAYERS
FOR THEIR TEAMS. ELLEN IS ALWAYS CHOSEN
FIRST. CHRIS IS ALWAYS CHOSEN SECOND. IF
ELLEN AND CHRIS ARE NEVER CAPTAINS, HOW OFTEN
DO THEY PLAY ON THE SAME TEAM?

Answer_____

Conceptual
Problems

7.

HOW MANY DIAGONALS DOES SQUARE WXYZ
HAVE?

Answer_____

8. THIS IS ONE UNIT:

WHAT IS THE VOLUME OF THIS SOLID?

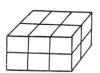

ANSWER_____

Fig. 1. Examples of types of problems on the mathematics achievement test

of teacher-directed whole-group instruction on predominantly low-level mathematics content followed by teacher monitoring of individual student seatwork that emphasizes mathematical knowledge and skills. The question is: Would these classroom processes be likely to facilitate the development of higher-level thinking skills in mathematics? To answer this question, we turn to some findings from recent research.

Classroom Processes That Facilitate Higher-Order Learning in Mathematics

Most reviewers of teaching effectiveness research, including the most recent ones (Brophy & Good, 1986; Rosenshine & Stevens, 1984), have concluded that direct instruction is the most effective instructional model for promoting the achievement of basic skills by students in elementary school. According to these reviewers, in direct instruction the teacher structures the information and presents it to the students, helps them to relate it to what they already know, monitors students' performance systematically, and provides systematic corrective feedback during recitation, drill, and practice activities. The teacher has the central role in the classroom and monitors classroom activities and manages student behavior and student academic work so that a high level of student engagement in the academic tasks in maintained.

Originally, reviewers cautiously asserted that direct instruction was effective for increasing student achievement of lower-level skills in reading and mathematics, but that the effects on higher-level skills were less certain (Rosenshine, 1979). More recently, however, reviewers appear to be advocating direct instruction for teaching higher-level skills in reading and mathematics also. For example, Brophy and Good (1986) concluded that the research shows that "students learn more efficiently when their teachers first structure new information for them and help them relate it to what they already know, then monitor their performance and provide corrective feedback during recitation, drill, practice, or application activity" (p. 130). They argued further that this applies to "any body of knowledge or set of skills that has been sufficiently well-organized and analyzed so it can be presented (explained, modeled) systematically and then practiced or applied during activities that call for student performance that can be evaluated for quality and (where incorrect or imperfect) given appropriate feedback" (p. 130). They suggested that this includes aspects of high-level as well as low-level mathematics.

Although direct instruction might be the most effective method for promoting students' achievement of lower-level skills in mathematics, it may be necessary but insufficient for enhancing achievement of higher-level skills. Higher-order thinking may require a less direct instructional approach that transfers some of the burden for teaching and learning from the teacher to the student and promotes greater student autonomy and independence in the teaching-learning process. Some tentative support for this hypothesis was provided from the results of a meta-analysis of studies of open and traditional teaching by Peterson (1979). She found that, although the effect sizes were small, they suggested that with the more direct approaches of traditional teaching students tended to perform slightly better on achieve-ment tests, but they did worse on tests of more abstract thinking, such as creativity and problem solving. Conversely, with less direct, more open approaches students performed slightly worse on achievement tests, but they tended to do better on creativity and problem solving. Similarly, in a recent observational study in fourth-grade mathematics classrooms, Peterson and Fennema (1985) found that certain types of classroom activities were related to low-level mathematics achievement while others were related to high-level mathematics achievement. We return to these findings later in this paper.

Doyle (1983) argued that researchers must be more precise in thinking about the potential effects of direct instruction on students' learning. Based on a review of research in cognitive psychology, Doyle suggested that less direct instruction might be needed to teach students higher-level thinking in mathematics, which is based on meaning and understanding. According to Doyle, self-discovery is important for students to derive a sense of mean-ing from the academic tasks in mathematics. Academic activity should be structured on the basis of what is known about mathematics and how stu-dents process information in mathematics. However, the activities must be unstructured enough to allow students the opportunities to experience the content directly in order for them to derive generalizations and invent algo-rithms on their own.

Thus, three dimensions seem to be important to an analysis of classroom processes that facilitate higher-order learning. These are: (a) rote learning versus meaning and understanding; (b) teacher control and direction of student learning versus student autonomy and independence in learning; (c) teaching and learning of higher-level executive routines and learning strat-egies. We turn now to a discussion of these topics.

Rote Learning Versus Meaning and Understanding in Mathematics

Research in cognitive psychology is based on the premise that knowledge and understanding are constructed by each person (see, for example, Doyle, 1983). Thus, in mathematics classes students are actively constructing knowledge and understanding about concepts and skills in mathematics.

Children actively construct knowledge and understanding. Recent cognitive science research in mathematics on addition and subtraction has shown that contrary to popular notions, young children are relatively successful at ana-lyzing and solving simple addition and subtraction word problems by using their own informal modeling and counting strategies (Carpenter, Hiebert,

& Moser, 1983; Carpenter & Moser, 1982). This informal knowledge could provide a basis for the student to develop both mathematics concepts and skills and a meaningful understanding of mathematics. However, the traditional mathematics curriculum does not build systematically on children's informal knowledge or support its development. Similarly, the exclusive use of direct instruction by teachers in the early grades to teach mathematical algorithms and procedures may result in children learning by rote the mathematics algorithms and procedures, but not acquiring a true mathematical understanding. Children may not see the connection between their informal knowledge and the formal mathematics they are taught in school. This may result in the formal mathematics being reduced to meaningless symbol manipulation.

One implication of cognitive science research in mathematics is that classroom instruction in mathematics must emphasize meaning and understanding in the beginning elementary school grades as much as in the older elementary school years. One might argue that, to engender meaning and understanding of mathematics and to encourage the development of higher-order thinking in students, teachers should introduce word problems from the very beginning of mathematics instruction, and word problems should not only be integrated into the mathematics curriculum but should form the basis of it (Carpenter et al., 1983). Currently, this is not being done in classroom instruction in mathematics. For example, traditional mathematics curriculum has been based on the assumption that computational skills must be learned before children are taught to solve even simple word problems (Carpenter, Fennema, & Peterson, 1984). Children are not given word problems to solve until they are deemed to have mastered the necessary computational skills. Then the word problems are usually provided at the end of a mathematics unit, emphasizing the computational skills taught in the unit.

Classroom studies of students' understanding. The results of naturalistic studies of elementary classrooms also suggest that students may not be deriving meaning and understanding from the instruction that they are receiving (Anderson, 1981; Blumenfeld, Pintrich, Meece, & Wessels, 1982). When elementary students are interviewed about their academic tasks, such as seatwork in mathematics, they tend not to focus on the meaning of the content to be learned. Rather, they report that their goal is to get the task finished or completed (Anderson, 1981). Anderson (1981) suggested that the student's focus on getting the task finished rather than on the quality or content of the task may be particularly typical of the low-achieving student. Low-achieving students may develop and use strategies that contribute to content coverage, but they do not necessarily develop strategies that contribute to content mastery. Moreover, they may not develop cognitive strategies, problem-solving abilities, or learning-to-learn skills, such as diagnosing and monitoring their own understanding, nor might they develop executive routines or strategies for remediating their own lack of understanding.

In our own naturalistic studies of elementary mathematics classrooms we have found that students' abilities to diagnose and monitor their own understanding is an important predictor of their mathematics achievement. In an initial study, 5th-6th grade students and their teacher were videotaped while the students were being taught a lesson on probability (Peterson, Swing, Braverman, & Buss, 1982; Peterson & Swing, 1982). Following each lesson, after the students had completed their mathematics seatwork, they were interviewed using a stimulated-recall procedure in which they were shown themselves on videotape to "stimulate recall" of their thoughts during mathematics instruction. They were asked several questions including the following: "During the math lesson, did you understand the part of the lesson you just saw on videotape?" When asked about their understanding of seatwork problems, students were shown a copy of the seatwork problems without answers, and asked to describe and talk about problems that they did or did not understand. We found that, independent of student ability, students' reports of their understanding of the lesson were positively and significantly related to their achievement on the seatwork problems (Kendall tau = .31 for Day 1; tau = .36 for Day 2) and scores on the achievement test (tau = .28). In addition, students who did not provide detailed explanations of what and why they had trouble understanding tended to do more poorly on the seatwork problems (tau = −.27 for Day 2) and on the achievement test (tau = −.31).

These results, showing a relationship between students' reports of their own understanding and their mathematics achievement and between students' capabilities to determine what and why they did not understand and their mathematics achievement, are consistent with cognitive conceptions of the learning process. It is possible that students who can accurately judge their own understanding of a mathematics problem have processed relevant information until the product of their processing matches a well-formulated idea of what the solution should be. The correct assessment of the problem solution and their performance of the necessary information processing ensure high achievement. Poor mathematics achievement and reports of not understanding may result when students cannot determine precisely what it is they are trying to do and thus cannot identify and perform the steps necessary to accomplish the learning goal.

Ability to monitor one's understanding. We replicated these findings for understanding in a second study in which fifth-grade students were learning a unit on measurement (Peterson, Swing, Stark, & Waas, 1984). Students were interviewed using a stimulated-recall interview following each lesson on measurement. Students' reports of understanding as assessed during the stimulated-recall interview were significantly positively related to students' mathematics seatwork scores (tau = .25) and to students' scores on the achievement test (tau = .24). Perhaps of more interest, however, and consistent with the viewpoint that specific student cognitions are the important mediating variable in classroom learning, our results showed that students who were able to provide a good explanation of what mathematics problem or part of the lesson they did not understand and why, tended to have significantly higher scores on the mathematics achievement test (tau = .24).

Students' provision of a good explanation for their lack of understanding was also significantly related to: (a) students' reported use of specific cognitive strategies; (b) students' reports of mathematics concepts and operations; and (c) students' reported checking of their answers against an external criterion, such as the answer provided by the teacher, or checking back over their own work. For example, in the following example, Jamie reported the process of diagnosing and monitoring his own mathematics answer by checking against the external criterion provided by the teacher:

Jamie: "When he [the teacher] was saying, you know, like he was asking people like 'what would number 7 be?' I'm am [sic] saying it like he says 'what's number 7?,' I say, 'It's approximately 10 meters'; or I'm saying, 'It will be 20 meters'; in my head and I know they're saying, '20 meters.' And he [the teacher] would say if it was right so I would see if I'm doing it right."

In addition to the correlations noted above, students' provision of a good explanation for their lack of mathematical understanding was also significantly related to their attitude toward mathematics. Taken together, this consistent pattern of correlations suggests that the reporting of a definitive or diagnostic reason for not understanding may reflect students' general tendencies to be involved actively in the mathematics task, regardless of their immediate understanding. Possibly as a result of their involvement, students constructed sufficient mathematical knowledge and understanding (as shown by their ability to report mathematics content) enabling them to diagnose precisely their difficulty, and then at some point to eliminate that difficulty with a minimum of new information. Moreover, the significant positive correlation between mathematics attitude and "a good explanation of lack of understanding" indicates the importance of motivational variables in actively processing information. Thus, students with positive attitudes may persist longer in the face of poor understanding than their counterparts with negative attitudes. Such persistence would increase their chances for eventual understanding. We turn now to a discussion of persistence, independence, and autonomous learning behavior as a variable that is importantly related to students' higher-order learning in mathematics.

*Teacher Control/Direction Versus Student Autonomy/Independence
and Persistence*

Fennema & Peterson (1985) have argued that autonomous learning behavior and independence are critical for the development of higher-level thinking in mathematics. On high-level mathematics tasks, it is not clear what the student must do to complete the task. (See for example, the problems in Figure 1.) For example, in a low-level computation problem a student can look at the addition, subtraction, multiplication, or division sign and know immediately what mathematical procedure must be performed to solve the problem. On the other hand, on a high-level mathematics problem the student first must figure out how to solve the problem. This requires independent thinking on the part of the student.

Other researchers have documented the importance of autonomy and independence in higher-level achievement in mathematics. Grieb and Easley (1984) argued strongly that independence in learning is essential to learning the conceptual framework of mathematics that enables one to continue to use mathematics or to do high-level cognitive tasks. Similarly, biographies of and interviews with outstanding mathematicians have shown clearly that these individuals had the traits of independence and persistence (Helson, 1980; Gustin, 1982).

Factors related to autonomous learning behavior. Fennema and Peterson (1985) hypothesized that the development of the kind of autonomy and independence needed to solve higher-level problems in mathematics might be particularly a problem for girls. This may be a partial cause for sex-related differences in mathematics achievement, which begin as early as fourth grade, with girls already doing less well than boys on higher-level problems in mathematics.

To test our hypothesis, we conducted a study in 36 fourth-grade mathematics classrooms in Wisconsin (Peterson & Fennema, 1985; Fennema & Peterson, 1986). In December, and again in May, students in the classes completed a mathematics test containing low cognitive level and high cognitive level items selected from the National Assessment of Educational Progress items as defined in Figure 1. During January through April, observers observed six randomly selected girls and six randomly selected boys in each class. Using a 60-second time sampling procedure, the observers recorded whether or not the students were engaged in mathematics activities and some characteristics of those activities. They also observed each teacher interacting with a target student and categorized the interaction on several dimensions. These included the cognitive level of the interaction (low-level or high-level mathematics), whether the teacher accepted a student's answer as correct, and the teacher's reaction to a "correct," "incorrect," "partially correct," or "nonevaluated" student response.[1]

Dependence as counterproductive for high-level learning. The results of our observational study supported our hypothesis. Girls' engagement in social activities, engagement in one-to-one interaction with a teacher, and receiving of help from the teacher were found to be *negatively* related to girls' higher-level mathematics achievement. We suggested that these classroom processes reflect a dependence on others that may be counterproductive for the kind of independent thinking that is needed to solve a high-level mathematics problem.

Some additional results for teacher feedback also seem to relate to students' dependence/independence. We hypothesized that teacher feedback about the mathematics strategy would be particularly important and effective for encouraging students' high-level mathematics responses and for increasing high-level mathematics achievement. Our results supported our hypothesis. Focusing on the mathematics strategy rather than on the answer was particularly important following a girl's high-level mathematics response and for encouraging girls' high-level achievement. Teacher prompting of the mathematics strategy was significantly positively related to girls' mathematics achievement. Prompting was coded whenever the teacher asked the student to continue or elaborate his or her response and provided a clue or hint to the student about what mathematics strategy should be used to solve the mathematics problem. Focusing on the strategy may be effective because most high-level mathematics responses have more than one right strategy that can be used to obtain the answer. Thus, the student needs to learn to focus on the strategy to solve the high-level problem.

The above findings suggest that persistence and other motivational aspects of autonomous learning behavior may be related importantly to students' mathematics achievement, particularly to higher-level mathematics achievement. We turn now to a discussion of some findings from our stimulated-recall interviews that suggest that persistence is importantly related to achievement, but only if the persistence is aimed at higher-order learning.

Persistence, motivation, and higher-order thinking. In our interviews with fifth-grade students about their thinking during mathematics instruction, we found that two categories of "affective" thoughts were significantly related to students' achievement and attitudes and to students' reported use of specific thinking strategies. These were motivational self-thoughts and reports of thoughts indicating that the student was interested in just "getting the task done."

Motivational self-thoughts were coded whenever a student's response suggested that she or he was putting forth special effort toward understanding the lesson or completing a problem, was enthusiastically involved in learning the lesson, or was being particularly conscientious or careful. Examples of motivational self-thoughts were the following:

> I *wanted* to know how to do it, but I didn't wanna um, goof around; I wanted

to get it so like if we had to use it, like later on the test; and we *did* have to use it later on the test, so I'm glad I did pay attention there.

Peterson et al. (1982) found that, independent of students' initial attitudes toward mathematics, students who reported motivational self-thoughts tended to have more positive attitudes toward mathematics at the end of the instructional unit than students who did not report such motivational self-thoughts (see Peterson & Swing, 1982). Moreover, students who reported such motivational self-thoughts also tended to report engaging in higher-order thinking strategies during mathematics instruction (Peterson et al., 1982). We obtained similar findings in our second study (Peterson et al., 1984). In our second study, we also included a second category of affective thoughts from the stimulated-recall interview which was defined as students' reports of thinking that suggested that the student was interested in just getting the task *done*. An example of this category is the student who said that he was thinking that he wanted "to finish it on time." Other examples are "I was thinking that I just wanted to get it (the lesson) over with," and "I was thinking about when it was going to be over." This category of affective thoughts was included because of results of a recent descriptive study by Anderson (1981) described earlier. In our study, we found that students' reports of "wanting to get done" were significantly *negatively* related to students' scores on the attitude pretest, students' seat-work scores, and students' scores on the attitude posttest.

Summary. In sum, the above results suggest that students' self-directed motivation and persistence are positively related to students' higher-level mathematics achievement. More specifically, students' self-direction and persistence should be aimed at deriving meaning and understanding from the mathematics task rather than on just "getting the task done" and at developing independent thinking skills and strategies for solving mathematics problems rather than on obtaining the "one right answer" to the mathematics problem. Further, teachers may need to provide opportunities for developing students' independent thinking, self-direction, and autonomy by encouraging students to solve mathematics problems on their own without relying on the teacher and by encouraging students' learning of divergent strategies for solving mathematics problems. We turn now to a discussion of higher-level executive processes and strategies for mathematics learning and to an examination of whether students seem to be learning these higher-order strategies in contemporary elementary mathematics classrooms.

Teaching Higher-Level Executive Processes and Strategies for Mathematics Learning

In our interviews with students in elementary mathematics classes, we also determined the extent to which elementary students report spontaneously using sophisticated kinds of executive routines and learning strategies and the extent to which students' reported use of these strategies was significantly related to their learning and achievement. In particular, we hypothesized that students' use of specific cognitive strategies would be significantly positively related to students' achievement. In contrast, students' reported use of more general cognitive processes might not be significantly related to achievement. Specific cognitive processes included checking answers, applying information at a specific level, reworking problems, rereading directions or problems, relating new information to prior knowledge, asking for help, using aides, using memory strategies, and trying to understand the mathematics lesson or do a mathematics problem by using a specific operation. Many of these processes and strategies have been identified as effective by educational researchers who have studied cognitive strategies. (See, for example, Kirby, 1984; Pressley, 1986; Pressley & Levin, 1983a, 1983b; and Weinstein & Mayer, 1986.) In contrast, more general cognitive processes included perceiving, processing, remembering, and producing a product.

Students' reports of specific cognitive strategies. In our two studies of fifth- and sixth-grade students' learning in mathematics classes, we found that in support of our hypotheses, the total number of general cognitive strategies reported by students was *negatively* related to students' mathematics achievement, but that the total number of specific cognitive strategies reported was *positively* related to achievement.

Several examples will serve to illustrate students' use of specific cognitive strategies and their mathematics learning. One example of checking is provided by the verbal protocol of Jamie given earlier. Students' reported cognitive processes often included statements about (a) a sequence of specific or general problem-solving steps that were used; (b) strategies required to work the problem; or (c) insights about the nature of the mathematics task. Examples of this kind of specific cognitive strategy included the following:

> I was thinking I had to measure both sides, and then times them together, and then make them to the centimeter square.

And in describing how to find the perimeter, another student reported:

> Sort of seeing the figure in my head, and adding it in my head, and thinking what "km" meant—trying to figure that out.

Students' reported use of cognitive strategies also sometimes included reports of mathematical operations that reflected complex thinking—comparison, decision-making, and consideration of alternatives. An example of such a report is the following statement by a student:

> I was kind of looking at my ruler off to the side, looking at it and then looking at the paper trying to guess which side. What? How close I could get?

In our research, we found that students report engaging in a wide variety of cognitive processes and strategies during mathematics instruction. Interestingly, however, those processes and strategies that students report most often were not those that are frequently proposed and researched by educational psychologists as facilitative of learning and achievement. For example, students seldom reported spontaneously using sophisticated kinds of learning strategies such as memory strategies, strategies for relating new information to prior knowledge, for discriminating, and for comparing information. To the extent that students' reports may be viewed as evidence of what elementary students do naturally to help themselves learn during mathematics, students' reports may be viewed as indicative of their knowledge about the learning process, that is, their cognitive knowledge. If, however, students' reports are viewed as indicating how little they know about cognitive processes and strategies that facilitate learning, then such lack of knowledge of cognitive strategies and metacognitive processes may, in fact, be limiting their potential for mathematics learning, and particularly for higher-order mathematics learning. Thus, these findings may be used as an argument for the need to change elementary mathematics teaching and elementary classroom processes so that they facilitate the development of higher-order executive routines and cognitive strategies. We will return to this issue of explicit teaching of higher-order executive strategies in our discussion on how we might change mathematics teaching and classrooms to facilitate higher-order thinking in mathematics.

Toward Effective Teaching for Higher-Order Thinking in Mathematics

Thus far in this paper, we have argued that the teaching and classroom processes that are occurring in elementary mathematics classrooms today may not be facilitating students' higher-level learning in mathematics. Further, we have identified several dimensions of teaching in classroom processes that may be importantly related to higher-level learning in mathematics. We turn now to a discussion of how elementary mathematics teaching in classrooms might be changed to promote students' higher-order thinking in mathematics. Specifically, we describe three examples from our own research in which we have attempted to alter elementary mathematics teaching in classrooms to facilitate students' higher-order thinking and autonomous learning behavior in mathematics. These examples are: (a) using small-group cooperative learning techniques to foster students' autonomy and independence and to facilitate students' learning of cognitive strategies and metacognitive strategies; (b) explicit teaching of cognitive strategies and metacognitive strategies to elementary students in mathematics classrooms; and (c) teaching first-grade teachers about results of recent cognitive science research in mathematics so that they might change their teaching and curricula to build on students' informal knowledge in order for them to develop a meaningful understanding of mathematics.

*Using Small-Group Cooperative Learning as an Adjunct to
Whole-Class Instruction*

Over the last decade, a substantial research literature has accumulated on the effectiveness of various small-group learning techniques. (See, for example, Slavin, 1983, for a review of this research.) In our research we have found that small-group cooperative learning may be an effective adjunct to whole-class instruction in the traditional elementary mathematics classroom (Peterson & Janicki, 1979; Peterson, Janicki & Swing, 1981; Peterson, Wilkinson, Spinelli, & Swing, 1984). For example, we have adapted small-group cooperative learning techniques to the format of mathematics instruction typically used in elementary classrooms by having students work together in small cooperative groups on seatwork problems. In this approach, the classroom teacher teaches the day's mathematics lesson for approximately 20 minutes, and then students work together on their mathematics seatwork assignments in small mixed-ability groups of four students. We have found that the positive effects on mathematics achievement seem to depend on the task-related interaction that occurs in the small group. What happens is that students learn by explaining an answer or explaining why an answer is incorrect to another student or by helping other students with their work. Each student works on his or her own mathematics seatwork, but when a student has a problem, another student helps. Research indicates that the students learn by explaining why the answer is incorrect and by helping the student come to see the correct answer. In addition, the receiver of the explanation may benefit from receiving an explanation that describes the kinds of strategies and processes that a student should use to solve the problem. (See, for example, Webb, 1983, and Peterson, Wilkinson, Spinelli, & Swing 1984.) However, we have also found that elementary students, particularly primary grade students, typically need some specific training in group interaction skills to work effectively in small groups (Peterson et al., 1984).

Active modeling of cognitive and metacognitive strategies. If one takes the perspective of the student as needing to develop cognitive strategies for higher-order learning in mathematics, one might argue that students learn effectively in small cooperative groups because they become actively involved in learning rather than passively receiving information being presented to them by the teacher. The following example of second- and third-grade students working together in a cooperative mathematics group on their seatwork presents such a picture of active learning. In this example, the small-group members were told to check their answers with one another after doing several seatwork problems. Johnny, the high-ability student in the group, is convinced by the lower-ability students in the group that his answer is incorrect:

Katie: [reading the answer from her paper] Dollar sign zero point forty-four.

Johnny: What? Whaddya mean "zero point forty-four?"

Katie: [pointing at Johnny's paper] Zero point forty-four.

Johnny: What? Eight nickels and four pennies equal thirty-six.

Katie: Eight nickels.

Johnny: Eight nickels. Eight times four equals thirty-two. Thirty-two plus four equals thirty-six.

Anne: No, it's forty-four, Johnny.

Katie: Let's go on with it.

Johnny: Which one are we on?

Anne: We're on five.

Katie: Five.

Johnny: [to Anne] Whaddya mean forty-four?

Anne: It's the eight nickels—forty-four.

Johnny: Ah, yeah. Wait a minute. Wait a minute.

Anne: It's forty-five.

Johnny: No, wait, it's not even thirty or forty-four. Naw, God, it's forty-nine.

Katie: Yeah.

Johnny: Forty-nine. No, wait a minute, it's forty-eight?

Anne: It's forty-four.

Johnny: It's forty-eight. Eight times . . .

Katie: Okay. [counting on fingers] 0, 5, 10, 15, 20, 25, 30, 35, 40, 41, 42, 43, 44.

Johnny: No, wait, wait a minute. Okay, okay, eight . . .

Anne: [counts on fingers to show Johnny] 5, 10, 15, 20, 25, 30, 35, 40, 1, 2, 3, 4.

Johnny: 5, 10, 15, 20, 25, 30, 35, 40. Okay, 40 + 4 = 44.

In the above example Johnny is convinced, or *has* to be convinced, that his answer is incorrect. The thinking-aloud process of the other students in the group makes their thought processes explicit to Johnny to convince him that his answer is wrong. Also, Johnny himself must think aloud and go through the problem-solving steps on his own before he is convinced that 0.44 is indeed the correct answer. One might hypothesize that not only are students learning the correct mathematics answer from such small-group interaction, but they are also more likely to learn the different strategies for arriving at answers to mathematics problems as well as possible skills and strategies for diagnosing and monitoring their own mathematics learning. In addition, small-group learning may foster students' independence and autonomous learning behavior because the implicit message is that students can learn on their own without the teacher present.

In sum, research suggests that elementary students might benefit both in terms of learning basic skills in mathematics and in terms of learning higher-order and metacognitive strategies and skills from interacting with their peers. We move now to a discussion of the effects on higher-order mathematics learning of the explicit teaching of cognitive and metacognitive strategies to elementary students.

Explicit Teaching of Cognitive and Metacognitive Strategies

Recent research on training cognitive strategies has demonstrated that cognitive strategy training can significantly enhance students' learning. More specifically, psychologists have found beneficial effects of training students in the following kinds of cognitive strategies: memory strategies, elaboration strategies, comprehension monitoring, self questioning, rehearsal strategies, planning and goal setting, comprehension strategies, verbal self-instruction and self-regulation, problem-solving strategies, hypothesis generation, and study skills (Kirby, 1984; Pressley & Levin, 1983a, 1983b; Weinstein & Mayer, 1986). Unfortunately, most cognitive strategy research has been conducted in laboratory settings with learning outcome measures that are narrowly defined and related directly to the experimental task (Peterson & Swing, 1983). Further support for the need for applications of cognitive strategy training to classrooms comes from our previous research in which we interviewed elementary students about the cognitive strategies that they employ during classroom mathematics instruction. In these studies, we found consistently that elementary students do not spontaneously use the sophisticated kinds of strategies that have been identified by psychologists as effective (Peterson et al., 1982; Peterson et al., 1984). These findings led us to conduct a study in which we trained elementary mathematics teachers to teach students to use thinking skills in their mathematics learning (Peterson, Swing, & Stoiber, 1986).

Fourth-grade teachers ($N = 30$) were assigned randomly to one of two interventions: (1) training in "academic learning time" and how to increase students' engaged time in mathematics; or (2) training to promote students' use of four cognitive strategies (thinking skills) in learning mathematics— (a) defining and describing; (b) comparing; (c) thinking of reasons; and (d) summarizing. As a part of the learning time intervention, teachers were instructed to use short learning activities in mathematics to soak up wasted time; to minimize interruptions; to use incentive systems; and to do active monitoring during mathematics seatwork. In the thinking skills group, teachers were instructed to teach their students to conceptually define and/ or pictorially represent mathematics terms, to compare, to evaluate procedures and answers, and to summarize. During February, teachers received their respective training through workshop sessions, feedback sessions, and printed materials. Each teacher was observed for several days prior to training and for eight days during March, April, and May to obtain a measure of post-treatment teacher behavior and classroom processes during mathematics class. In May, students were interviewed about the thought processes and cognitive strategies that they engaged in during a selected mathematics lesson. In December, and again in May, students completed a mathematics achievement test based on the categories shown in Figure 1 and using released and parallel items from NAEP (1979).

Regression analyses of posttest achievement on pretest scores showed significant ability-by-treatment interactions between classes for high-level mathematics achievement, conceptual mathematics achievement, and achievement on word problems. High-level mathematics thinking, conceptual thinking, and word problems were targeted in the thinking- skills intervention. The pattern produced by the ability-by-treatment interactions showed that higher-ability classes achieved more on the high-level cognitive mathematics items, the conceptual items, and word problems in the thinking-skills intervention than in the learning time intervention. In contrast, lower-ability classes achieved more on these mathematics problems in the learning-time intervention than in the thinking skills intervention. Additionally, within-class regression analyses showed significant ability-by-treatment interactions for high level mathematics achievement, word problem achievement, and mathematics achievement total. Within classes, the lower-ability students benefited more from the thinking skills intervention than from the learning time intervention.

The between-class and the within-class ability-by-treatment interaction results taken together suggest that a fairly high level of average mathematics ability for a given class was necessary for a teacher to implement the thinking skills treatment effectively with that class. However, the observation data suggested that, once the teacher implemented the thinking skills intervention effectively, then lower-ability students within that class tended to benefit more than the higher-ability students. Perhaps lower-ability students within a class were helped by the thinking skills treatment more than higher-ability students because the thinking skills treatment "compensated" for or remediated aptitude processes or cognitive strategies that lower-ability students did not naturally have. The thinking skills intervention thus provided the lower-ability students with strategies and processes to use in solving mathematics problems and learning mathematics. Acquisition of these strategies then permitted them to learn as effectively as the higher-ability students within the class who had already naturally developed the strategies on their own. Verbal protocol data supported this hypothesis.

Effects of strategy training on the low-achieving student. To provide a concrete example of how a low-ability student within a time class differed in his reported cognitive processes from a low-ability student within a thinking skills class, we present two verbal protocols, the first one for a student in the time group and the second one for a student in the thinking skills group.

A low-ability student in the time class was given the following classic "van" problem that he had not seen before and was asked to think aloud as he worked the problem:

29 students went on a field trip. Each van could hold 8 students. How many vans are needed?

The following verbal protocol was obtained from the student in the time group:

Student: You have to multiply 8 x 9 um, um, then you have to multiply 9 and 7. Then you have to multiply um, 8 and 2.

Interviewer: Tell me everything that you're thinking.

Student: Um, then you have to multiply 7 and 2. Then you add 2 um, and you add 6 and 3. Then you add 6 and 1.

Interviewer: Ok, are you done? Yes? What do you think the final answer is?

Student: 14,992

Interviewer: Ok, what were you thinking about besides the problem?

Student: Um, the answer.

To illustrate the use of the thinking skills by a lower-ability student in a thinking skills class, we present an excerpt from the student's retrospective seatwork protocol. The student was given the following problem that had appeared on the student's seatwork earlier:

Sunset trail is 4 miles longer than Lonesome trail. Lonesome trail is 13 miles long. How long is Sunset trail?

Interviewer: Tell me everything you were thinking or doing to figure out the problem, so I can learn to think about it just like you.

Student: Well, I was thinking that if Lonesome trail was 13 miles long, if I added to 13, I would get how long 4 miles was. And if I added a different number with 4, I wouldn't have got the same number as I got to see how long it was. And if I added it this way 13 + 4, and then I added 4 + 3 is 7, and I put the 1 down here and I got 17 miles long.

Interviewer: What things that you learned in math class either today or some other time were you thinking about or doing?

Student: Well, when I add my problem I always think what my way is: Should I add this or times it? And I found out that I added it. And when I was done I also found out that my answer was right because there was one part, and I had a whole, and I had a part, and I needed another part, and this was the part to make the whole.

The first segment of the above student's protocol indicates extensive use of thinking of reasons. The second part of the student's verbal protocol demonstrates use of the thinking skill of comparing. Contrasting the verbal protocol of this lower-ability thinking skills student with the lower-ability time student, one finds greater use of cognitive strategies as well as a more thoughtful approach to the problem that was being solved. For example, the time student came up with an answer that was absurd given the small numbers in the problem.

Summary. These qualitative data provide evidence of the effectiveness of thinking skills instruction in altering the way that low achieving students approached, solved, and thought about word problems in mathematics. The verbal protocol data suggest that the thinking skills treatment had either a "compensatory effect" or a "remedial effect" for lower-ability students within a class by providing them with cognitive strategies that they did not have naturally. Use of these thinking skills by both lower-ability and high-ability students thus decreased the effect of mathematics ability on students' achievement within the thinking skills classes. In contrast, in the time classes, without such a cognitive intervention to compensate for the lower-ability students' lack of cognitive strategies, a strong positive within-class relationship existed between students' prior mathematics achievement and students' posttest mathematics achievement.

Using Results from Cognitive Science Research to Improve Classroom Instruction

In a four-year research project funded by the National Science Foundation, Thomas Carpenter, Elizabeth Fennema, and I are attempting to provide first-grade teachers with recent findings from cognitive science research in children's mathematics learning so that teachers alter their first-grade mathematics instruction and curricula to facilitate higher-order learning in mathematics. As we discussed above, recent research on cognitive science in mathematics suggests that children's informal knowledge of addition and subtraction could provide a basis for the teacher to develop students' mathematical knowledge, skills, and concepts as well as a meaningful understanding of mathematics. Teachers might modify the traditional mathematics curriculum to build systematically on children's informal knowledge. Teachers might introduce mathematics word problems early on in first-grade to form the basis of the curriculum. We refer to this new instructional approach as "Cognitively-Guided Instruction." (For a more complete description of our research project, see Carpenter, Fennema, Peterson, 1984; Fennema, Carpenter, Peterson, 1986; and Carpenter, 1987.)

Research on Cognitively-Guided Instruction. Cognitively-Guided Instruction (CGI) is based on several principles. First, teachers need to be familiar with research findings on children's solution of addition and subtraction problems. They need to learn to classify addition and subtraction word problems and to identify the processes that children usually use to solve these problems. Second, teachers need to be able to assess not only whether a child can solve a particular word problem but also how the child solves the problem. Within a CGI classroom, a teacher analyzes children's thinking by asking appropriate questions and listening to children's responses. Teachers' analyses and understanding of children's responses are based on a framework derived from research on children's thinking. Third, teachers need to use the knowledge that they derive from assessment and diagnosis of the children to design appropriate instruction. This will probably mean changing the regular traditional first-grade curricula to include greater emphasis on problem solving and particularly on solving of word problems. In addition, teachers may need to alter normal instructional practices of whole-class instruction and individual seatwork to facilitate students' more active learning by including the use of learning centers and small, cooperative peer groups. Teachers might also encourage a student to work a problem aloud for the whole class or to tell other students how she or he solved a problem. Finally, an underlying principle is that teachers should organize instruction to involve children so that they actively construct their own knowledge with understanding. Further, teachers should ensure that elementary mathematics instruction stresses relationships between mathematics concepts, skills, and problem solving.

We are conducting a year-long study with 43 first-grade teachers to determine whether teachers who participated in our workshop on Cognitively-Guided Instruction have altered significantly their curricula and their mathematics instruction in classroom processes to build on the principles of Cognitively-Guided Instruction. Further, we are interested in whether these classroom processes and teaching techniques differ from the control teachers' classroom processes and teaching techniques and, if so, whether as a result of altering their teaching and classroom instruction the Cognitively-Guided Instruction teachers will be more effective in facilitating their children's higher-order learning in mathematics, particularly children's learning of addition and subtraction word problems and meaningful understanding of first-grade mathematics. Moreover, through our systematic observations of teachers' classrooms as well as through interviews with teachers and students in the classrooms, we hope to understand and explain those teaching processes and classroom processes that relate to first-grade students' higher-order mathematics learning.

Conclusions and Implications

In this paper we have argued that the challenge for educators in the next decade is to improve students' learning of higher-order skills in mathematics. Further, we have suggested that recent research and theory suggest that the following classroom processes might facilitate higher-order learning in mathematics: (a) a focus on meaning and understanding mathematics and on the learning task; (b) encouragement of student autonomy, independence, self-direction and persistence in learning; and (c) teaching of higher-level cognitive processes and strategies. Findings from naturalistic studies of classrooms suggest that teachers do *not* currently emphasize the above in their teaching of elementary mathematics. Thus, the above analysis might serve as a framework both to guide future research and to improve educational practice.

Research on Cause-Effect Relationships Between Classroom Processes and Students' Higher-Order Learning

Many questions remain to be answered. A fundamental question is: For each of the above three processes, to what extent does a cause-effect relationship exist between the process and students' higher-order achievement? For example, does a greater teaching focus on meaning and understanding lead to greater higher-order learning in mathematics? If so, what should be the relative emphasis on rote memory of facts and on cognitive skills STET in mathematics? More concretely, in teaching addition and subtraction in first-grade to what extent and when should teachers emphasize student learning of addition number facts to a level of "automaticity" as opposed to emphasizing meaningful understanding of addition and subtraction concepts and problems? And, when and how should teachers incorporate number fact learning with concept learning and mathematical problem solving? Moreover, what are the resulting effects on students' fact learning, higher-order learning and problem solving in addition and subtraction?

Similar specific questions need to be addressed for the other processes discussed. For example, does teaching a specific cognitive strategy enhance a students' higher-order mathematics learning? Numerous strategies might be studied and their effects on mathematical learning examined. These include goal-specific strategies, monitoring strategies, higher-order (planning and organizing) strategies, specific mathematics strategy knowledge, and general meta-strategy knowledge. (See, for example, Pressley, 1986.) Also, researchers need to investigate the effects of coordination of strategies and knowledge on students' mathematics learning (Pressley, 1986).

Research on How to Change Teaching

In addition to demonstrating that the above processes *do* significantly affect students' higher-order mathematics learning, researchers need to determine *how* teaching and classrooms might best be changed to facilitate these processes. Possible approaches to change range on a continuum from a more global "educational" approach to a more specific directed "training" approach.

An example of the former approach is the one that we have taken in our study of Cognitively-Guided Instruction (CGI). CGI involves an increased instructional emphasis on all three processes described above as follows:

(a) Emphasis on teaching for meaning and understanding of addition and subtraction concepts and problems as integrated with and building toward the learning of number facts.

(b) A focus on first-grade students as active learners and "constructors" of knowledge who have a large informal knowledge base in mathematics when they enter school. Thus, the role of the teacher becomes one of facilitating active student learning by diagnosing and understanding students' current levels of strategic thinking and organizing instruction to encourage students' movement to the next higher cognitive level.

(c) Encouraging the teaching and learning of specific strategy knowledge, such as using "counting on" or "derived facts" to solve addition and subtraction word problems as well as verbalization of these strategies by children. Teaching of such strategies might occur through direct instruction by the teacher but also might be done through less direct methods such as guided)discovery or peer modeling. (See Pressley, Snyder, & Carigila-Bull, in press, for a good discussion of alternative modes of strategy instruction.)

In sum, our CGI approach constitutes an attempt to intervene on all three dimensions of classroom instruction above. As a result, it is a full-barreled approach rather than a targeted approach. Our rationale was that, if we obtain effects with a more global approach, then we will investigate the effects of specific components of the approach in later research.

In addition, CGI teachers were *not* trained in specific techniques for altering their classrooms and curricula. Thus, in employing a CGI approach to "educating" the teachers and working with them as "thoughtful professionals," who construct their own knowledge and understanding, we remained consistent with our theoretical framework drawn from cognitive science. However, the result will undoubtedly be large variations in the extent and degree to which CGI teachers change their curricula and teaching to incorporate CGI principles. Indeed, our preliminary observations at CGI teachers' classroom implementation suggested that this is the case.

In contrast, one can imagine a more targeted approach to changing teaching that would involve translating each of the three dimensions given above into specific teaching behaviors, teaching strategies, and curriculum changes that teachers might implement. Teachers might be directly instructed to implement these skills and changes. Direct instruction might include providing teachers with lesson plans, scripts, verbal protocols, and curriculum materials that will permit the teachers to implement these changes easily with minimal teacher development effort and decision making. A potential problem with this latter "training" approach is that research has shown that such "teacher-proof" approaches often fail if they do not consider and build on teachers' implicit theories and beliefs and teachers' existing knowledge structures (Clark & Peterson, 1986).

In sum, research is needed to examine the question of *how* teachers and classrooms might best be changed to focus on and emphasize the classroom processes that have been shown to facilitate higher-order thinking in mathematics. Researchers will need to consider the distinction between a teacher "education" approach and the potential effects of various approaches along the continuum from "education" at one end to "training" at the other.

A coordinated, concerted effort is required to change teaching to facilitate higher-order learning in mathematics. As researchers and educators, the challenge is before us for the next decade.

NOTE

[1]We obtained the following estimates of inter-rater reliability for observations on these categories: student engagement in high-level mathematics activities (.95); student engagement in low-level mathematics activities (.97); high-level mathematics teacher-student interaction (.80); low-level mathematics teacher-student interaction (.86); teacher accepts a student's mathematics answer as correct (.92); teacher gives neutral feedback to students' mathematics answer (.80); teacher gives answer (.73); and teacher prompts mathematics answer (.76).

REFERENCES

Anderson, L. M. (1981). *Student responses to seatwork: Implications for the study of students' cognitive processing* (Research Series No. 102). East Lansing, MI: The Institute for Research on Teaching, Michigan State University.

Bennett, W. J. (1986). *First lessons: A report on elementary education in America.* Washington, DC: U.S. Department of Education.

Blumenfeld, P. C., Pintrich, P. R., Meece, J., & Wessels, K. (1982). The formation and role of self-perceptions of ability in elementary classrooms. *Elementary School Journal, 82,* 400-420.

Brophy, J. E., & Good, T. L. (1986). Teacher behavior and student achievement. In M. Wittrock (Ed.), *Handbook of research on teaching,* (3rd ed., pp. 328-375). New York: Macmillan.

Carpenter, T. P. (1987, January). *Teaching as problem solving.* Paper presented at the Conference on Teaching and Evaluating of Problem Solving, San Diego, CA.

Carpenter, T. P., Fennema, E., & Peterson, P. L. (1984). *Cognitively guided instruction: Studies of the application of cognitive and instructional science to mathematics curriculum.* Proposal funded by the National Science Foundation, Washington, DC.

Carpenter, T. P., Hiebert, J., & Moser, J. M. (1983). The effect of instruction on children's solutions of addition and subtraction problems. *Educational Studies in Mathematics, 14,* 56-72.

Carpenter, T. P., & Moser, J. M. (1982). The acquisition of addition and subtraction concepts. In R. Lesh & M. Landau (Eds.), *The acquisition of mathematical concepts and processes* (pp. 7-14). New York: Academic Press.

Clark, C. M., & Peterson, P. L. (1986). Teachers' thought processes. In M. C. Wittrock (Ed.), *Handbook of research on teaching* (3rd ed., pp. 255-296). New York: Macmillan.

Doyle, W. (1983). Academic work. *Review of Educational Research, 53,* 159-200.

Fennema, E., Carpenter, T. P., & Peterson, P. L. (1986). *Studies of the application of cognitive and instructional sciences to mathematics instruction* (Technical Progress Report, August 1, 1985 to July 31, 1986). Madison, WI: Wisconsin Center for Educational Research, University of Wisconsin-Madison.

Fennema, E., & Peterson, P. L. (1985). Autonomous learning behavior: A possible explanation of gender-related differences in mathematics. In L. C. Wilkinson & C. B. Marrett (Eds.), *Gender influences in classroom interaction* (pp. 17-35). Orlando, FL: Academic Press.

Fennema, E., & Peterson, P. L. (1986). Teacher-student interactions and sex-related differences in learning mathematics. *Teaching and Teacher Education, 2*(1), 19-42.

Goodlad, J. L. (1984). *A place called school.* New York: McGraw-Hill.

Grieb, H., & Easley, J. (1984). A primary school impediment to mathematical equity: Case studies in role-dependent socialization. In M. Steinkamp & M. L. Maehr, *Women in science* (pp. 317-362). Greenwich, CT: JAI Press.

Gustin, W. (1982). *Learning to become a mathematician: The development of independence.* Paper presented at the annual meeting of the American Educational Research Association, New York.

Helson, R. (1980). The creative woman mathematician. In L. H. Fox, L. Brody, & D. Tobin (Eds.), *Women and the mathematical mystique.* Baltimore: Johns Hopkins University Press.

Kirby, J. R. (Ed.). (1984). *Cognitive strategies and educational performance.* Orlando, FL: Academic Press.

National Commission on Excellence in Education. (1983). *A nation at risk: The imperative for educational reform.* Washington, DC: U.S. Government Printing Office.

National Assessment of Education Progress (1979, May). *The second national assessment: 1977-1978: Released exercise set.* Denver, CO: Author.

Peterson, P. L. (1979). Direct instruction reconsidered. In P. L. Peterson & H. J. Walberg (Eds.). *Research on teaching: Concepts, findings and implications* (pp. 57-69). Berkeley, CA: McCutchan Publishing Corporation.

Peterson, P., & Fennema, E. (1985). Effective teaching, student engagement in classroom activities, and sex-related differences in learning mathematics. *American Educational Research Journal, 22*(3), 309-335.

Peterson, P. L., & Janicki, T. C. (1979). Individual characteristics and children's learning in large-group and small-group approaches. *Journal of Educational Psychology, 71,* 677-687.

Peterson, P. L., Janicki, T. C., & Swing, S. R. (1981). Ability x treatment interaction effects on children's learning in large-group and small-group approaches. *American Educational Research Journal, 18,* 453-473.

Peterson, P. L., & Swing, S. R. (1982). Beyond time on task: Students' reports of their thought processes during classroom instruction. *Elementary School Journal, 82,* 481-491.

Peterson, P. L., & Swing, S. R. (1983). Problems in classroom implementation of cognitive strategy instruction. In G. M. Pressley & J. R. Levin (Eds.), *Cognitive strategy research*: *Educational applications* (pp. 267-287). New York: Springer-Verlag.

Peterson, P. L., Swing, S. R., Braverman, M. T., & Buss, R. (1982). Students' aptitudes and their reports of cognitive processes during direct instruction. *Journal of Educational Psychology, 74,* 535-547.

Peterson, P. L., Swing, S. R., Stark, K. D., & Waas, G. A. (1984). Students' cognitions and time on task during mathematics instruction. *American Educational Research Journal, 21,* 487-515.

Peterson, P. L., Swing, S. R., & Stoiber, K. C. (1986). *Learning time vs. thinking skills: Alternative perspectives on the effects of two instructional interventions* (Program Report 86-6). Madison, WI: Wisconsin Center for Education Research, the University of Wisconsin-Madison.

Peterson, P. L., Wilkinson, L. C., Spinelli, F., & Swing, S. R. (1984). Merging the process-product and the sociolinguistic paradigms: Research on small-group processes. In P. L. Peterson, L. C. Wilkinson, & M. Hallinan (Eds.), *The social context of instruction: Group organization and group processes* (pp. 126-152). Orlando, FL: Academic Press.

Porter, A. C., Floden, R. E., Freeman, D. J., Schmidt, W. H., & Schwille, J. R. (1987, April). *A curriculum out of balance: Elementary* school mathematics. *Paper presented at the annual meeting of the American Educational Research Association, Washington, DC.*

Pressley, M. (1986). The relevance of the good strategy user model to the teaching of mathematics. *Educational Psychologist, 21*(1,2), 139-161.

Pressley, M., & Levin, J. R. (Eds.). (1983a). *Cognitive strategy research: Psychological foundations.* New York: Springer-Verlag.

Pressley, M. & Levin, J. R. (Eds.). (1983b). *Cognitive strategy research: Educational applications.* New York: Springer-Verlag.

Pressley, M., Snyder, B. L., & Cariglia-Bull, T. (in press). How can good strategy use be taught to children?: Evaluation of six alternative approaches. In S. Cormier & J. Hagman

(Eds.), *Transfer of learning: Contemporary research and applications*. Orlando, FL: Academic Press.

Romberg, T. A., & Carpenter, T. A. (1986). Research on teaching and learning mathematics: Two disciplines of scientific inquiry. In M. C. Wittrock (Ed.), *Handbook of research on teaching* (3rd ed., pp. 850-873). New York: Macmillan.

Rosenshine, B. V. (1979). Content, time, and direct instruction. In P. L. Peterson & H. J. Walberg (Eds.), *Research on teaching: Concepts, findings, and implications* (pp. 28-56). Berkeley, CA: McCutchan.

Rosenshine, B., & Stevens, R. (1984). Classroom instruction in reading. In P. D. Pearson (Ed.), *Handbook of reading research* (pp. 745-798). New York: Longman.

Slavin, R. (1983). *Cooperative learning*. New York: Longman.

U.S. Office of Education. (1987, January). *Japanese education today*. Washington, DC: Author.

Webb, N. M. (1983). Predicting learning from student interaction. Defining the interaction variables. *Educational Psychologist, 18*, 33-41.

Weinstein, C. F., & Mayer, R. F. (1986). The teaching of learning strategies. In M. C. Wittrock (Ed.), *Handbook of research on teaching* (3rd ed., pp. 297-314). New York: Macmillan.

ACKNOWLEDGEMENTS

A version of this chapter was presented at the "Research Agenda Conference on Effective Mathematics Teaching," University of Missouri, Columbia, MO, March 11-14, 1987 and at a conference on "Motivation, Schooling, and Intellectual Development," University of Michigan, Ann Arbor, MI, January 30, 1987. Work on this paper was supported in part by a grant from the National Science Foundation (Grant No. MDR 8550236). Some research reported in this paper was supported in part by grants from the U.S. Department of Education (Grant No. NIE-G-81-0009 and Grant No. NIE-G-84-0008) and from the National Science Foundation (Grant No. SED 8109077). The opinions expressed in this paper do not necessarily reflect the position, policy, or endorsement of the U. S. Department of Education or the National Science Foundation.

Interaction, Construction, and Knowledge: Alternative perspectives for mathematics education[1]

Heinrich Bauersfeld,

Universitat Bielefeld, FRG

| "Whatever we see could be other than it is. Whatever we can describe at all could be other than it is. There is no priori order of things." | "Alles, was wir sehen, konnte auch anders sein. Alles, was wir uberhaupt beschreiben konnen, konnte auch anders sein. Es gibt keine Ordnung der Dinge a priori." |

Ludwig Wittgenstein, *Tractatus Logico-Philosophicus* No. 5.634

1. The Theoretical Autism

Following Kuhn's (1970) remark that paradigmatic change in scientific communities announces itself through the discussing and questioning of the paradigm current, this ought to be an interesting period of research in mathematics education. Indeed, new research approaches are developing and promising new fields have opened up, many of these by challenge from computer use. But, as far as I can see, we are not realizing the challenges arising from new perspectives. We do not pursue the consequences systematically; we do not think them through.

What does it mean, that there seems no prospect of arriving at universal or overall theory of teaching-learning processes that we will all accept? How then do we, as researchers, deal with a plurality of competing, incompatible, or contradictory theories? We also know that research does not produce sufficient or even an overall orientation for the mathematics teacher's activities. How do teachers in their daily practice cope with these blanks? And which orientation can we as observers reconstruct from a teacher's observed and documented actions? Interestingly, sociologists now investigate how scientists develop and make decisions about scientific statements (Anselm Strauss, Fritz Schuetze, Karin Knorr, et al.), though, unfortunately, not yet in the field of mathematics education.

It may be useful, at least for a moment, to look at our own profession critically and from interdisciplinary and philosophical points of view. Researchers, perhaps, are the captives of preferred "language games," (Wittgenstein, 1953) just as teachers are, though certainly of different language games. In comparison with teachers, perhaps researchers do have "other" knowledge about classroom realities, rather than "better" knowledge. Can we really repudiate the suggestion that our research work confounds the observance of standards and methods with the development of scientific knowledge?

It is not necessary to go as far as Lyotard did with his provocative analysis of the "knowledge of the postmodern era" (1979), or as Feyerabend's anar-

27

chic "anything goes" (1975). Also, I do not suggest a kind of self-sufficient abstract reflexion about our professional proceeding, which makes the millipede trip over his own feet. Yet, if we just consume the many novelties and do not make use of the potential contrasting power of the alternative perspectives in order to learn about the blinding self-evidence of our common conceptualizing and theory-building, then the reproduction of the traditional may prevail in the end.

As a preliminary basis for discussion a short account is given of how my own perspective has developed across the last two decades. Then, as an illustration of many challenging findings, I will present an example of our research work as developed in the group Krummheuer, Voigt, and Bauersfeld, and relate this to an alternative view on a few key concepts in mathematics education. Finally, some remarks on educational theories in general are added under the outlined alternative perspective: about what we may expect of it and how we may use it for developing the quality of the professional research work.

At a conference of consultants of the European Community in Brussels recently, Tim O'Shea from Edinburgh was asked about the presumable impact of computers and Artificial Intelligence (AI) development on education and on the schools in the nineties. His laconic answer was the counterquestion, "How do you want it?!", thus pointing at the potential power of resolute inventions rather than surrendering to fatalism. I share this confidence in human inquisitiveness, creativity, and responsibility.

2. The Development of a Perspective

It may be useful to describe the shift of research emphasis in general, and over the last decades, by giving a personal view. The following brief historical review serves to give a closer look at perspectives and their function in research.

It is more than twenty years ago now since I started the Frankfurter Projekt on New Maths in the Primary Grades with 42 teachers of "experimental" classes and another 40 teachers of "control" classes, who cooperated with us over four years from grade 1 through 4, and with about 3000 children. At that time Hilgard's *Theories of Learning* (third edition 1966) was in favour, as well as Bernstein's distinction between "restricted" and "elaborated codes" and their relation to social retardation, to "language barrier," and to learning difficulties (1971). The main focus was on the *student*, his or her learning prerequisites and abilities, traits and attitudes.

The principal interest of the project was with the possible effects of a New Maths curriculum on students' thought and language. We therefore developed our own materials, textbooks, and teacher's manuals, trained the "experimental" teachers, and assisted their classroom work, whilst the separate evaluation team, headed by my late colleague Valentin Weis, collected a rich bouquet of empirical data from the two groups of students and their teachers. In the end, we arrived at a new curriculum and at many insights, about ourselves, about mega-dollar projects, about school innovation, and about the teaching and learning of mathematics, especially.[2]

Over the project's long period of life, the flow of findings and insights led to critical reflexion and to changes of the project perspectives. The conflict with the initial design was answered provisionally through adding more variables to it. A striking early outcome of the "formative" evaluation (as it was called) was a much larger variation among the classes from within each project group compared with the variation between the experimental group and the control group. Though at first sight disappointing for the developers of the material, the finding drew attention to the teacher's key role in classroom processes. Consequently, and supported by many other outcomes, the perspective shifted from the student to the *teacher*, to his prerequisites, attitudes, motivation, readiness for and engagement in innovation, and so forth. Again this was thoroughly in accordance with the main focus of educational research at that time (e.g., Kounin, 1970; the reviews in Good & Brophy, 1973; and Brophy & Good, 1974).

A dissident from the evaluation team, E. Wolff, tired of the statistical analyses, produced intensive interviews with selected project teachers (up to 7 hours length), to a certain extent under a psychoanalytic perspective. From that, we learned a lot about these teachers' subjective balance between personal identity and the perceived professional role, about the function of participation and joint responsibility in the project procedures, and so forth, and we developed a better understanding for certain bewildering difficulties in the relation between project team and teachers.[3] Since then, questions about how teachers and students perceive themselves, about their relations to each other and to their mutual image of one another, and about how these interpretations influence the activities in the mathematics classroom have developed a peculiar dynamic for us. As a result, the evaluation team has split up into a larger group pursuing the initial statistical (quantitative) design and a smaller "soft" group working qualitatively on case studies.

Another call for "soft" methods arose from the troubles we had with analyzing the detailed audio-tape records from systematic observations of small-group work of children in first grade classes in 1969-70. Without any intervention, teacher students recorded each mathematics lesson with the same group of children over three months. From the informal analyses we learned about the relative symmetry of classroom actions: Both teacher and students contribute to the classroom processes. It is a jointly emerging "reality" rather than a systematic proceeding produced or caused by independent subjects' actions. As one would describe this perspective now:

Teacher and students jointly constitute the reality of the classroom (Mehan, 1979); they form a social sub-group, or function as a "system" (Seeger, 1987); or they "create the culture" of the classroom (Bruner, 1986).

The team members of the "hard" evaluation line tried to complement their numerical analyses through many additional small-scale investigations, which were conducted in parallel to the large data collections (e.g., on rigidity and strategies in problem solving, creativity related to different New Math areas, spatial reasoning, etc.). Somewhat outstanding in mathematics education, although again in accordance with a main wave of educational research at that time, was the application of "cognitive style" dimensions like "reflectivity versus impulsivity" (Kagan) and "field-dependence vs. field-independence" (Witkin), not only to students and teachers as individuals, but to the effects of a match or mismatch of these dimensions in *teacher-student relationships*. From that point of view, the relations between a highly impulsive teacher and a highly reflective student appeared to be the worst case possible in classroom "interaction" with reference to the teaching and learning of mathematics (Radatz, 1976). Related to individual differences of students, our interest was with a deeper understanding of how teachers perceive and cope with students' actual errors. As well as a more general questioning of the concept of "error" (a closer discussion will follow), we considered a joint generation or causation of the phenomenon of error, but had no descriptive means or methods for investigating it. In those days, educational psychology used "interaction" in the narrow statistical sense of an interaction between variables, such as trait-x-treatment or trait-x-trait interactions. This use of the word "interaction" did not meet our expectations in trying to understand individualizing teaching (but it is typical for the strong tie which mathematics education had and still has with educational psychology). It was years later that I became aware of independent developments in *sociology* where interaction is used with more fundamental meanings, as in "symbolic interactionism" (Mead, 1934; Blumer, 1969; Goffman, 1974), "ethnomethodology" (Garfinkel, 1967; Mehan, 1975, and 1979; Coulter, 1979), "sociolinguistics" (Cazden, 1972; Cicourel, 1974), and, to some extent, "cross-cultural studies" (Cole et al., 1974; Cole & Means, 1981).

In the end, the pattern of the developing research processes can be characterized by the shift of main perspectives across the decade of the project: from subject matter structures and the related learning of students, to the teacher's teaching and its conditions, and then to the teacher-student relationship from a psychological point of view (later, and after the termination of the project, to the social interaction of teacher and students). For the active researcher, each position arrived at relativizes the preceding perspective. This can lead him to disregard previous perspectives, and even to a relative blindness caused by the deliberate shift of focus (as many actual computer-related developments demonstrate, to the disadvantage of the intended aims). And there is no easy integration as conclusion, either for the researcher or for the interested teacher.

The analogy to the development of research in education in general may be conspicuous. But, clearly, this would be too limited a view. The "didactical triad" of *subject matter-student-teacher* has ruled over research in mathematics education for too long. Through tradition and habituation these three main perspectives and the related descriptive systems have gained a captivating power in the research community.[4] Thus, it has become quite difficult to withstand its eloquent temptation. Subject matter language, for instance, the language of mathematics, is widely used for the description of students' concepts as well as for teachers' intentions and for syllabi. Adequate descriptions and understanding of students' actual conceptualizing suffer for that reason. Also, the common decomposition of classroom events into teachers' actions, students' actions, and conditions, and where each person's action "causes" another person's reaction, as transmitted through the many descriptions and current ascriptions of causes, that are at hand for interpretations, blind the eyes of the researcher to interactive patterns and the intersubjective constitution of norms for actions (cf.Mehan, 1975).

We have arrived at an important point: The related problem is of a theoretical nature and by no means one of *façon de parler* only. Theories, in the common understanding, sets of hypotheses that are used as long as inconsistencies, contradictions, or falsifications do not appear, imply the danger of going round in circles. The ongoing repair can lead to a state of perfection of the theory's core, where there is no falsification beyond. And, more, this continuous use acts (for the user) like a closed system of related meaning ("abgeschlossener Sinn-bereich," A. Schuetz, 1979), inside which self-evidence, reliability, and certitude accompany and reinforce the use of concepts and hypotheses; they are just "at hand." Moreover, what is true with mathematics applies to most of the theories: their related technical language is not made for a reflexion of the theory from inside itself. The network of concepts is not self-reflexive; the theory is not its own meta-theory.[5]

Under a radical constructivist's perspective there is, in the end, no escape from this. But relative alienation and distancing can start from three different points: the language and models from other disciplines, cross-cultural approaches of ethnomethodology, and historical analyses of the transmutation of related concepts and contexts. Since it is the research policy of our institute, Institut fur Didacktick der Mathematik (IDM) to develop and to pursue alternative theoretical perspectives to cope with such fundamental theoretical problems, different orientations and approaches have been developed. For the following I will limit myself to the perspective under development cooperatively with Krummheuer, Voigt, and Bauersfeld.[6]

After the termination of the Frankfurter Projekt (1974), and with the outset of the institute (IDM) in Bielefeld, my own work separated step by step from the mainstream of research and development in mathematics education. Attempts at identifying and combining relevant factors or variables became less important and quantitative designs became completely unimportant. This is to say, we did not work quantitatively any longer, but, nevertheless, we continued to follow closely outcomes and reviews of related research. Using the experiences gained we began to investigate specific situations in the mathematics classroom, but under a completely different perspective: as social settings and under a micro-view. The techniques of videotaping and transcribing became crucial in order to make possible a repeated replay of every situation or scene. This opened the way for a deliberate shifting of the actual focus of attention from replay to replay, for the application of different descriptive means and "language games," and, thus, for the arrival at different interpretations—different, in particular, from the familiar prima facie explanations. The researcher then finds himself forced into the reflexion and comparison of the applied bases of knowledge and the presumed fit with the perceived context. The attempts towards holistic interpretations get into troubles in many aspects, for which no adequate theories of analysis have been developed so far.[7] Inevitably, therefore, the interpretation of classroom scenes, and of teaching-learning problems in particular are cases not only of an application of scientific theories but also of a creative use of everyday understanding and shared (sub-) cultural knowledge. This, evidently, increases the urgency of a reflexive control of the research procedures in order for the arrival at certain coherent theoretical frameworks and for the development of standards for such microethnographical analyses (Voigt, 1984; Bauersfeld, 1982). Typical for these research approaches is the "abductive" generation of hypotheses (Peirce, cited in Voigt, 1984), but I will not go deeper into methods here. The next sections will present details of the joint interactionistic and constructivistic perspective related to concrete cases.

3. Patterns of Interaction in the Classroom

One of the early episodes analyzed was taken from a mathematics lesson that was part of a radio broadcast "Experiences with School Reform" (Germany, 1976). We will use the episode here as an illustration for the results from the analyses of many more school scenes and for some conclusions related to possible theoretical backgrounds. The radio program was made in order to demonstrate the importance of individualized teaching and of adequately differentiated interventions of the teacher. In addition to the audio-tape and the school textbook there was also available a transcript of the lesson, an interpretation of the lesson from educational scientists (but not mathematics education specialists), and a reaction from the teacher to the lesson (published in Diederich & Lingelbach, 1977, pp. 50f.).

The lesson dealt with the following problem from a widely used textbook which, for the moment, each student had to solve at his desk:

"A medicinal spring gives 200 hectoliters per hour. How big is the outcome of the spring *a)* daily, *b)* in one month, *c)* annually?

The episode covers the few minutes when the teacher (T) "intended to help a student (S), who obviously could not go on alone."

(1) *T:* . . . there is no special month given,

(2)   then you take thirty days and figure with the thirty days,

(3)   and in *a)* then we are going to have the quantity of water for one day already . . .

(4)   and how much then will this be for one month?

(5) *S:* (keeps silent)

(6) *T:* Now, you know, one month has thirty days . . .

(7) *S:* (affirmatively) . . . mhm . . .

(8) *T:* . . . and now?

(9) *S:* (keeps silent)

(10) *T:* One hour . . . you need not to say at present how much a day has, that you must figure out then,

(12)   well now, *one* day has *x* hectoliter, OK?

(13)   And then you must take the *x* hectoliter *how many times?*

(14) *S:* (keeps silent)

(15) *T:* Come on, how many did we say for one month?

(16) *S:* Thirty days.

(17) *T:* There you are, *thirty times x hectoliter,*

(18)   that will give the hectoliters for one month.[8]

3.1 *Preliminary Overview.* Under the professional view of *school mathematics*, an observer may find the discussion goes about the proceeding from part *a* to part *b* of the task. This step of the solution requires an inference to be made from the quantity of water in one day to the quantity in thirty days: In thirty days you will get thirty times as much as in one day. This multiplicative structure has to be recognized, like the second step in the rule of three. Since the orientation for the solution, which the teacher tries to provide for, is given for the whole task as an overview, the solution of subtask *a* is only hypothetically available; it is not yet accomplished. This will add to the difficulties the student apparently has with the required conclusion. So the teacher gives remedial information (note more at each stage), leaving time for the student's reaction in each case, expecting him to join the demonstrated analysis. Since the student does not come to grasp the adequate mathematical operation, it is the teacher who, in the end, presents the verbal solution.

3.2 *A possible interactionistic reconstruction.* A closer reconstruction under an *interactionistic* perspective leads to the following turn-by-turn interpretation of the episode:

(1) The teacher opens the episode applying a general routine (strategy): If the student does not know how to proceed with a given question, then recollect and clear the data and leave the construction of the procedure, the operation and so forth, to the student. This is what the teacher performs in lines 1-3 (abbreviated L.1-3) with a tiny, inviting pause at the close. The expected reaction fails to come. The student does not respond. So the teacher continues with reshaping the question (L.4). With "and" and "then" he lays some emphasis on the connection between "one day" to "one month". Apparently the teacher wants to hear an answer like "thirty times" or something similar.

(2) To the student's silence (L.5) the teacher now reacts with a less pretentious intervention. He reduces his expectations for the student's achievable reactions. The teacher narrows his intervention to the crucial role of the factor *thirty:* He repeats the premise "one month has *thirty* days" (L.6), adding a slight but distinct stress to the word "thirty," thus labelling it as a keyword. The friendly "now, you know" serves to transmit the encouraging assumption of something like: "I know, you can do that, if you make an effort . . ."

(3) From his experiences with mathematics lessons and from many remedial teacher-student discussions, nearly every fourth grader will understand an accentuated number word in a repeated utterance from the teacher as: You have to operate with this number now. (If you have just used the number yourself, it is wrong and you have to correct it now.) Surely, the teacher knows this as well, though more in a subconscious manner. So the student's mere affirmative sound cannot satisfy the teacher's expectation. There is no progress towards the expected answer. His "and now?" therefore sounds more demanding, sharper, and less friendly.

(4) For the student this will signal that the plot thickens. (A look at his face through a videotape replay would be helpful for the interpretation here.) The student keeps silence. Surely this is difficult for him, considering the strong obligation for action in this institutionalized setting: He is next to move in the turn-taking game. But he resists (L.9). This has a confusing effect on the teacher, as the scrappy start of his next move demonstrates (L.10). His initial strategy did not work so far. The student's difficulties are more complicated,

obviously. And the time passes by; he has to come to a positive close. So the teacher finds himself under pressure as well. Setting out for a clearer, more profound explanation of the relation between "one day" and "one month," he tries to clear the hypothetical status of the first factor in the multiplication he has in mind, the quantity of water per day (L.10-11). Pushing aside the difficulty, he introduces the notion of "x hectoliter" (L.12) and the stressed "one" again points at the crucial relation. He prepares to put things straight: He gives himself a push initially ("well now") and he appeals to the student's consent ("OK?"). He therefore reformulates the problem, using the new notion, and comes out with a distinct question. The question, however, again makes a lower demand on the student's capability, because the student is expected simply to fill in the void at the end of the question as in a completion test. The teacher's expectation and effort now has come down to a fight just for one word (L.13).

(5) By saying "thirty" or "thirty times" the student would surely pass the minimal threshold which the teacher has set for him, and the episode would terminate. But the student keeps silence (L.14). The teacher's confidence decreases near to zero from this. He no longer applies recognized strategies which build on the student's active construction and leave space for discovery and self-guided control. From general hints for the developing of insight with the student, he has narrowed his interventions step by step until he arrives at a plain "recitation game" (Hoetker & Ahlbrand, 1969) now, in order at least to make the student say the required word. Opposite to the quality of his expectations, the teacher's tension has reached a certain climax. The somewhat condescending "come on," opening his question, carries quite clear connotations for the student, like, "This at least you will have to know" and "if that fails, I will have to give up".

(6) The final delivery of the keyword (L.16), by the way, the very first "substantial" contribution of the student, leads to a sudden decline of the teacher's accumulated tension. In his relaxation and his final pedagogical satisfaction ("there you are") the teacher recasts and completes the student's answer, as he realizes his utterance (L.17). But, in fact, the teacher produces the expected solution himself: With emphasis, he says, "thirty times x hectoliter". And relieved he sends an additional explanation after it (L.18), like an emotional full stop of the episode. Surely the student will relax also, but he is left unable to solve the task in this case as the subsequent scenes prove.

3.3 *Patterns of interaction.* Stepping back from the concrete case and comparing it with other episodes, we can identify patterns of the related activities, something like an underlying grammar of the process. It is *as if* the participants follow hidden regulations they are not aware of. Teacher and student(s) jointly produce such *patterns of interaction* by means of certain routines. In the present case one may think of a "funnel" as an appropriate metaphor. Since we do not discuss methods here, I will limit the presentation to a short description of the "funnel pattern of interaction" here:

- The teacher recognizes a student with difficulties; for example, the student made an error, cannot draw an inference, cannot cope with a task and so forth.

- The teacher opens with a short question in order to stimulate self-correction. He receives an unsatisfactory reaction.

- The teacher then goes further back to collect and clear prerequisites for the insight, aiming at an "adequate" reaction from the student. *Adequate* at this stage is already an approximate fit with the teacher's expectation.

- Continued deviant answering on the student's side meets on the teacher's side a growing concentration on the stimulation of the "adequate" answer through more precise, that is, narrower, questions. Thus the standard for "adequateness" deteriorates, the quality of the discussion decreases.

- Step by step the teacher, in fact, through what he does, reduces his presumption of the student's actual abilities and self- government in a way that is quite opposite to his intentions and in contradiction even to his subjective perception of his own action (he sees himself "providing for individual guidance").

- The student realizes both the simplified but stiffer demands and the growing tension. If his own constructive efforts do not match the teacher's narrowing objectives, his chances for a successful finish dwindle. (In a late stage the teacher sometimes rejects more detailed answers given to an earlier, broader question of his.) Thus the tension intensifies with teacher and student.

- When the deterioration has come down to a simplest exacting recitation or completion from the student, the culmination is reached. Just one expected word from the student then can bring the teacher to a presentation of the complete solution by himself.

- In every case, the pattern terminates with the presentation of the solution, independently from who produces it. Also, the pattern can break off at any intermediate state, but the later the stage the less probable this becomes.

To avoid possible misunderstandings, it may be useful to point out the special nature of such "patterns of interaction": Obviously the common principle of "cause and effect" is to a certain extent difficult to apply here. Neither the teacher nor the student alone can be blamed for the course of the process; they are not aware of the covert pattern of their joint actions. As a product from *social* interaction, the patterns develop from the mutual reflexive expectations and interpretations of the actors and their related moves, from the implicit "obligations for action," which are typical for the institutionalized educational processes, and from the teacher's and the student's routines as acquired across many shared classroom experiences. It is the network of relations among these constituents that produces the pattern of interaction. Open to some extent in its beginning, the pattern becomes stabilized from turn to turn, the freedom for choices dwindles, and there is restrictive power with the process in the end. Ethnomethodologists say: "Realities are always realities becoming" (Mehan, 1975, p. 203). Or, specialized to our case: Teacher and student(s) *constitute* the *reality* of the classroom *interactively*. This also makes clear, from theoretical explanation, why predictions in concrete classroom situations are so difficult and shaky. Through the analysis, reconstruction, and comparison of many cases, the patterns of interaction appear to function as a quite reliable hidden grammar of classroom activities.

What are the effects of these hidden regulations? In the special case of the "funnel pattern," teacher and student leave the situation relaxed, the teacher, in particular satisfied and convinced that he has "carefully reacted to the different difficulties of individual students" (Diederich & Lingelbach, 1977, p. 49). But what is left with the student? All in all, he has not manifested more than "mhm" and "thirty days." Even if he could solve the task subsequently, following the teacher's given recipe, we could not be sure about whether there is more than the mere accomplishing of a recipe.

As open for intelligent actions as the pattern starts, equally narrow and impoverished is the constituted meaning at the end. And, surely, this is not the type of an enriched environment that promotes insight and transmits the desired orientations on the meta-cognitive level.

3.4 *Conclusions*. In general, we can state: It is difficult to imagine a mathematics lesson without teachers' and students' routines, and, without some pattern of interaction, it would be hard to go through. These regularities reduce the complexity of the classroom processes. They make the institution bearable for teachers and students, and stabilize their mutual actions.

On the other hand, however, the desire to "keep it going", to keep the lesson smooth and trouble free, works essentially against the main aims of the mathematics syllabus. The covert social structure of the classroom actions masks or supersedes the mathematical structures, which the teacher has in mind and which he has tried to stage, and which the student can construct only through the regularities of his own (internal and external) actions. In these situations, the learner's adaptive efforts towards an acceptable use of mathematical symbols and language are bound to generate context- and problem-specific routines and skills rather than insight, self-confidence, flexible strategies, and autonomy. *The mathematical logic of an ideal teaching-learning process thus becomes replaced by the social logic of this type of interactions.* This, perhaps, is the core of the notorious school-generated failure of so many mathematical school careers.

At present we are investigating relations among patterns of interaction, certain traditions, and belief systems. Related to school practice, we cooperate with teachers from different types of schools in in-service training courses and similar settings. As a first step we present selected cases in a narrative mode, point at the pattern of interaction and the generating routines in these concrete cases, and discuss the possible effects for the student's learning with them. Anger and bewilderment usually bring about reasonable motivation and engagement for an ensuing joint development of constructive alternatives, preparations for related lessons, and so forth, in small-group work. We also use video-tapes, which the teachers themselves have produced.

Yet we do not know enough about the meaning teachers can develop from such approaches for their daily practice, nor about the stability of such changes. It is easier to say what we do not want. Certainly, we do not follow the diffuse, common recommendation for repair: "Raise teachers' reflexion," or: "Develop teachers' meta-knowledge," which was the final conclusion of so many research papers in the past years. There are too many instructional and principle difficulties with the development of meta-cognition, of context-free, general strategies for problem solving, of direct changes of collaterally learned orientations for action, and so forth, which suggest that the realization of such hopes will not be easy (Weinert & Kluwe, 1984; Bauersfeld, 1984; Schoenfeld, 1987). Knowledge from an instructional situation is not automatically transferred to a different situation of possible application. And, perhaps, in terms of memory structures there is no such thing as general knowledge, which can be retrieved context-free and in its presumed generality (Bauersfeld, 1985).

4. The Need for Perspectives Complementing One Another

> When and if we pass beyond the unspoken despair in which we are now living,
> . . . a new breed of developmental theory is likely to arise . . . I think that its
> central technical concern will be how to create in the young an appreciation of
> the fact that many worlds are possible, that meaning and reality are created and
> not discovered, that negotiation is the art of constructing new meanings by which
> individuals can regulate their relations with each other. It will not, I think, be an
> image of human development that locates all of the sources of change inside the
> individual, the solo child. For if we have learned anything from the dark passage
> in history through which we are now moving it is that man, surely, is not 'an
> island, entire of itself' but a part of the culture that he inherits and then recreates.
> The power to recreate reality, to reinvent culture, we will come to recognize, is
> where a theory of development must begin this discussion of mind. (Bruner, 1986,
> p. 149)

The general view of reality as a "reality created" by the subject (person), and not as a kind of a discovered objective reality, is a typical perspective of constructivists (e.g., Ernst von Glasersfeld, Paul Cobb, Jere Confrey, and others, including Bruner himself). For them, the subjective construction of reality leads to a viable adaptation to the resistance and to the obstacles of a world, which the subject can describe and understand only via these constructions, and more: A world, which the subject itself is part of. Because he is capable of producing concepts, relations and routines, the subject actively searches for regularities and meanings in repeating actions and situations. The adaptation functions in the sense and to the extent that the subject changes his constructions until relative, actual satisfaction has been achieved. Thus, the outcomes of the subjective constructions of realities are at best viable hypotheses, but not objective truth, about "the world." There is no cognition like a direct "reading" or a "discovering" of "the" world. Bruner's own dissociation from his earlier fascination with "discovery" is noteworthy for the perspective under discussion here.

On the other hand, the descriptive means and the models used in these subjective constructions are not arbitrary or retrievable from unlimited sources, as demonstrated through the unifying bonds of culture and language, through the intersubjectivity of socially shared knowledge among the members of social groups, and through the regulations of their related interactions. Apart from the adaptation to the resistance of "the world," learning is characterized by the subjective reconstruction of societal means and models through the negotiation of meaning in social interaction and in the course of related personal activities. New knowledge, then, is constituted and arises in the social interaction of members of a social group (culture), whose accomplishments reproduce as well as transmute the culture (e.g., of the mathematical community, of teacher and students of a class, etc.). The notion of "negotiation is the art of constructing new meanings" (Bruner, 1986, p.149) goes beyond the limited psychological focus on the dichotomous relations between student and subject matter, or between teacher and student (see note 4), and opens the field for sociological, interactionistic perspectives respectively (cf. Mead, 1934; Mehan, 1975, 1979; Miller, 1986).

Altogether, the subjective structures of knowledge, therefore, are subjective constructions functioning as viable models, which have been formed through adaptations to the resistance of "the world" and through negotiations in social interactions. This *triadic nature of human knowledge* makes impossible an ascription of causes, which would dissect internal from external causations (Seiler, 1984; Seiler & Wannenmacher, 1983). The separation for analytical purposes may be necessary, but is helpful only provided the researcher does not lose sight of the fundamental inseparability. There is good reason to assume that Piaget was aware already of this interrelation (Furth, 1983). It is in this sense that in our analyses the emphasis is laid upon the pursuit of the interactionistic perspective, because this part of the triadic nature appears to be the most underestimated and underdeveloped part in research in mathematics education.

5. Some Remarks on Theory

It may be helpful in explaining the perspective I have outlined, to close with some reflexions about what theories are good for and about consequences for future research work.

The Fundamental Relativism Scientific theories can be seen as models of realities. Models never map or match the reality meant in total. They serve their predictive purpose through the designing of relevant elements and relations, that is, through selection and neglect. To arrive at certain selections requires a perspective, a standpoint.

It is with the constitution of our world that it can become a reality given under a certain perspective only. Perspective is one of the components of relaity. . . . A reality with an identical face for every observer is nonsense. . . . Spinoza's species aeternitatis, the overall, absolute standpoint, does not exist."	Die kosmische Wirklichkeit ist so beschaffen, dass sie nur in einer bestimmten Perspektive zur Geltung kommen kann. Die Perspektive ist eine Komponente der Wirlichkeit. . . . Eine Wirklichkeit, die von allen Standpunkten gleich aussieht, ist ein Nonsens. . . . Spinozas species aeternitatis, den überalligen, absoluten Standpunkt gibt es nicht."

(Jose Ortega y Gasset, *"El tema de nuestro tiempo,"* 1923)

If one accepts this fundamental relativism, at least in the field of the human sciences, then he will question the objectivity of judging a theory with the means or under the perspective of a competing theory.

Language and Perspective Theories depend on language. They appear as descriptive systems, for example, as relatively closed networks of concepts and relations. Due to the "two-faced" nature of language, "that it serves the double function of being both a mode of communication and a medium for representing the world about which it is communicating" (Bruner, 1986, p. 131), the descriptive system, the language of the theory, has a potentiality to mold and restrict thinking through force of habituation. Since the language of a theory carries the perspective both as mode and as medium, an uncritical user need not be aware of the covert, implicit guidance (control) of his focus, in particular, when he is using parts of the theory only. For the same reason there is no plain transformation of one theory into another theory by translation of the language of one theory into the language of the other theory.[9]

Uncertainty Principle Since the perspective of a theory functions like a fixed focus, we can state an Uncertainty Principle for theories in education by analogy with the famous Heisenberg Principle: Similar to the physical categories of position and impulse of an electron, one cannot elaborate two incompatible perspectives with arbitrary precision. Concentrated and systematic pursuit of one perspective forces other perspectives into blindness, e.g. the focus on *what* is learned or taught versus how a person is learning or teaching it, or on the student's or teacher's *individual* actions versus the social interaction of both of them, and so forth. The need for coherent description and explanations (or interpretations) then makes the other perspective reappear as "noise," as "complementary issues," as "intervening variables," and so forth, which hamper the precision of description and weaken the predictive power of the theory.

Perspective and Object The usual research game of turning these disturbing interventions into main objects of follow-up investigations and, in effect, of extending theory this way, is doomed to circularity and failure. With the change of perspective, the descriptive system inevitably changes and the object is no longer the "same" object. Incompatible perspectives are inappropriate bases for an integrating theory; indeed, there is no chance at all, except "God's eye view" as Putnam said (quoted from von Glasersfeld, 1983). To speak of the "dual nature" of "the" object, therefore, is more likely an obscuring metaphor, which gives support to the idea of an homogeneous object (like "the" electron) behind the dichotomy.[10] The unstable "same"ness of the object across different technical languages also challenges the common metaphor of taking a "higher standpoint," a macroview, under which the dichotomies may disappear or may appear as incompatibilities, eliminable by means of a dialectical synthesis, or as parts of an enclosing "system", and so forth. Meaning and object change with the perspective; there is no adding of theories for greater accuracy.[11]

Historicity Scientific theories come into being mainly through opposition against prevailing theories. There is an ongoing transmutation of preferred perspectives, perceived problems, and actual solutions across history. So-called progress in education and in the human sciences in general thus can be construed as a permanent change of perspectives and of related descriptions and meanings. There is no accumulative growth of understanding, because in the human sciences "truth" is by nature historical truth. The search for universals and consistencies across historical periods, as well, is bound to an actual perspective and can produce an answer for the present only, which may not hold tomorrow.

Understanding Theories As opposed to the context-bound ascription of meaning in everyday language use, scientific theories are presumed to rest upon the strict use of their technical terms. Researchers often pick the labels (the words, the "signifiers") for their key categories following a contiguity relation between the concept (the "signified") they have in mind and one specific of the many facets of meaning ascribed to the word in everyday use. Functioning as both help and hindrance, this facilitates initial understanding and access to the theory, but also gives rise to the illusion of an easy meta-basis for criticism in both directions, from the theory directed outwards as well as against the theory from outside; whereas serious criticism of a theory would at least require an adequate understanding of the network character of its technical terms (see note 10). The construction of a metatheory capable of executing the critical comparison of competing theories will fail due to the impossibility of an uniting metaperspective and because of the (related) nonexistence of a universal language. How to proceed, then?

The Crucial Use By no means must this theoretical relativism lead us into resignation or despair. Rather, the considerations are meant for purifying the common beliefs in precision, systematics, generality, scientific truth, and so forth. In the end, it is the life in mathematics classrooms which processes and provides the evaluation of theories in mathematics education; though, again, theories are required to "read" these judgments. We never escape either from the Hermeneutic Circle[12] or from the need for theoretical orientation.

For research in mathematics education in the present state, it may be useful to take care of at least three neglected tasks:

- to clarify theoretical positions and their perspectives, in particular with reference to their implications for interpreting and orienting classroom activities. This may help to overcome significant limitations as produced through habitual professional perspectives (from which even recent AI developments suffer).

- to use and to develop alternative perspectives and related theoretical frameworks, since only competitive descriptions and contrasting issues have the power to produce challenge and to disquiet and break the customs of self-evident routines and explanations. There is no other way to distance oneself from habit and to allow for effective critical comparisons and reflexions.

- to investigate how theories function in practice (are "practiced"); for example, the use teachers make of them, the classroom practice or "culture" which may be related to them, but also how researchers themselves deal with theories and apply them to the analysis of classroom processes.[13] The reconstruction of documented classroom events may teach us more about the predictive power of a theory and about how it enables a more detailed understanding than analyses of the theory on an abstract level. One of the philosopher Lichtenberg's famous aphorisms points exactly at this pragmatical attitude.

| One must not judge people for their opinion, rather for what these opinions make of them. | Mann muss die Menschen nicht nach ihrer Meinung beurteilen, sondern nach dem, was diese Meinungen aus ihnen machen. |

(Lichtenberg, 1742-1799, *Aphorisms*)

NOTES

1. Invited paper for the Conference on Effective Mathematics Teaching in Columbia, Missouri, March 11-14, 1987, sponsored by the National Council of Teachers of Mathematics (NCTM).

2. The project was financed by Max-Traeger-Stiftung and Stiftung Volkswagenwerk. The curriculum has been published under the name "alef" with H. Schroedel, Hannover, and is still in use locally. The research reports have appeared in four volumes in a mimeographed series. For an English assessment see, for example, B. Moon (1985), *The New Math Curriculum*, Basingstoke: Taylor & Francis. (also Bauersfeld, 1972.)

3. Indeed,it is somewhat appalling to realize the degree to which many recent innovations disregard the knowledge about innovative processes in education from one or two decades ago (e.g., the many books from R. Havelock, D. Walker, L. Smith, L. Stenhouse, E. House, et al.) The neglect is spectacular indeed with the introduction of computers into schools and with the production of related software.

4. Within the classical triad Teacher-Student-Mathematics, the issue of "methods" in mathematics education had to explain the Student-Mathematics relation, the aspect of learning, and the issue of "classroom organisation" (Unterrichtsmethode) presented the *Teacher-Student* relation, the aspect of teaching, while the *Teacher-Mathematics* relation was a matter of teachers' academic preparation, and, so, an aspect of teacher training or further qualification. From a psychological view, each of the relations has been elaborated, essentially without transgressing the triad's borderlines. At present, two approaches go beyond the usual limits: The (social) interactionistic perspective on a mainly sociological basis and the epistemological perspective with a strong historical and philosophical background.

5. This, evidently, is the price paid for the more or less clear-cut definitions of technical concepts. Technical languages therefore are by nature not self-reflexive as opposed to the possible use and unique advantage of everyday language.

6. A larger group (Otte, Bromme, Jahnke, Seeger, Steinbring) works under an interdisciplinary, epistemological perspective with a certain relationship to "activity theory" (Vygotsky, Leontjev, et al.). They ask for the development of "knowledge," according to R. Thom's definition of "the real problem which confronts mathematics teaching", that is "the problem of the development of 'meaning,' of the 'existence' of mathematical objects" (quoted from Steinbring, 1986). A fundamental idea is the issue of "complementarity" which guides the identification and hierarchisation or interrelations between mutually contradictive aspects of a development (Otte, 1986). Concepts, therefore, are discussed as both label and procedure. Action and concept, consciousness and object, develop mutually, with a key role of the "means" used in these processes. Consciousness is seen as "concrete consciousness" in the sense of "being conscious of objects. " Therefore, the paradox statement holds: "The object is in the mind, and the mind is in the objects." (Otte, 1986; for more details see also Jahnke & Seeger, 1986; Seeger, 1987; and Bromme & Steinbring, 1987).

7. Sociological theories on discourse and interaction usually neglect aspects of content-specifity, which are just of interest for the mathematics educator, and of the special conditions in institutionalized settings. Concepts like "working consensus" (Krummheuer, 1983), "covert obligation for action" (Voigt, 1984, 1985) and "domain of subjective experiences" (Bauersfeld, 1983), just to name a few, have been shaped to meet these difficulties.

8. Since the original of the scene is in German, I have to apologize for a rather stiff and non-colloquial translation, possibly. But, I hope, this will not affect its function as illustrative material for the underlying pattern.

9. The many attempts for a translation of concepts from Behavioristic, Piagetian, Psychoanalytic, Cognitive, and other approaches into each other may serve as examples. AI's new ploy is to use the language of Cognitive Psychology and other disciplines as metaphors for machine

procedures: The computer "learns," "interacts," "reacts intelligently," etc. The confounding descriptions transmit the illusion of sameness, identifying man and machine.

10. In a critical analysis of the relations between Wittgenstein's notion of "language game" and psychoanalysis, A. Lorenzer points to the shift of meaning, which the very same vocabulary undergoes through the use in different theories. His conclusion: "One has to bring to mind all concepts and statements from within the related theoretical frameworks"; if not, "even insignificant, innocuous concepts are to be confused." (Lorenzer, 1977, p. 20)

11. Interestingly, the relative better fit of a theory must not lead to the total refutation of other theories. For economical and other reasons, they may survive in limited areas of application, as is true with classical physical science. All of present space flights and aeronautics are based upon non-relativistic physics, as far as we know.

12. W. Humboldt and W. Dilthey formed the notion. The Hermeneutic Circle marks the fact that all understanding is pre-structured in the sense that it rests upon preceding processes of understanding, but is not necessarily determined by those. The social sciences have taught us that in socially constituted realities, "facts" themselves are accomplishments of understandings. Descriptions of classroom events, therefore, can be seen as pre-structured interpretations of pre-structured interpretations.

13. There are interesting, recent sociological investigations done in research institutions and laboratories on how researchers present their conjectures, how they arrive at their decisions about the validity of statements and about publication, how they structure their related negotiations, and so forth. (Anselm Strauss, Fritz Schuetze, Karin Knorr, et al.)

"One misconception of the traditional paradigm is that the big problems of education can be solved by research."

(J. S. Allender, 1986, p. 187)

REFERENCES*

Allender, J. S. (1986). Educational research: A personal and social process. *Review of Educational Research*, *56*, 173-193. (Overview of "nontraditional" approaches, useful for comparative reading.)

Bauersfeld, H. (1972). Einige Bemerkungen zum Frankfurter Projekt und zum "alef"-Programm. In Arbeitskreis Grundschule (Ed.), *Materialien zum Mathematikunterricht in der Grundschule* (pp. 237- 246). Frankfurt/Main: Arbeitskreis Grundschule e. V.

Bauersfeld, H. (1982). Analysen zur Kommunikation im Mathematikunterricht. In H. Bauersfeld, H. W. Heymann, G. Krummheuer, J. H. Lorenz, and V. Reiss (Eds.) *Analysen zum Unterrichtshandeln* (pp. 1-40). Koln: Aulis Verlag Deubner & CoKG.

Bauersfeld, H. (1983). Subjektive Erfahrungsbereiche als Grundlage einer Interaktionstheorie des Mathematiklernens und -lehrens. In H. Bauersfeld, H. Bussmann, G. Krummheuer, J. H. Lorenz, and J. Voigt (Eds.), *Lernen und Lehren von Mathematik* (pp. 1-56). Koln: Aulis Verlag Deubner & CoKG.

Bauersfeld, H. (1984). Ergebnisse und Probleme von Mikroanalysen mathematischen Unterrichts. In W. Dorfler & R. Fischer (Eds.), *Empirische Untersuchungen zum Lehren und Lernen von Mathematik*. (pp. 7-25). Wien: Holder-Pichler-Tempsky.

Bauersfeld, H. (1985). Contributions to a fundamental theory of mathematics learning and teaching. *Proceedings of the Annual Meeting of the Canadian Mathematics Education Study Group*. Quebec, Canada.

Begle, E. (1979). *Critical variables in mathematics education*. Washington, D.C.: Mathematical Association of America and National Council of Teachers of Mathematics. (The last great attempt to organize "a substantially complete list of all the variables which have been studied for their effects on mathematics education." p. XVII)

Bernstein, B. (1971). *Class, Codes and Control*. London: Routledge & Kegan Paul. (since 1961, many articles before this systematic account)

Blumer, H. (1969). *Symbolic Interactionism*. Englewood Cliffs, NJ: Prentice Hall.

---

*Professor Bauersfeld has provided some references, with commentary, as suggestions for further reading.—Eds.

Bromme, R. & Steinbring, H. (1987). *Die epistemologische Struktur mathematischen Wissens im UnterrichtsprozeB. Eine empirische Analyse von vier Unterrichtsstunden in der Sekundarstufe I.* Unpublished manuscript, Bielefeld: Institut fur Didaktik der Mathematik, Universitaet Bielefeld.

Brophy, J. E. & Good, T. L. (1974). *Teacher-student relationships.* New York: Holt, Rinehart & Winston.

Bruner, J. (1986). *Actual minds, possible worlds.* Cambridge, MA: Harvard University Press.

Cazden, C., John, V., & Hymes, D. (1972). *Functions of language in the classroom.* New York: Teachers College Press, Columbia University.

Chomsky, N. (1986). *Knowledge and language.* New York: Praeger.

Churchland, P. (1984). *Matter and consciousness.* Cambridge, MA: M.I.T. Press.

Cicourel, A. V., Jennings, K. H., Jennings, S. H. M., Leiter, K. W. C., MacKay, R., Mehan, H., and Roth, D. H.(Eds.) (1974). *Language use and school performance.* New York: Academic Press.

Cole, M. & Scribner, S. (1974). *Culture and thought.* New York: Wiley & Sons.

Cole, M. & Means, B. (1981). *Comparative studies of how people think.* Cambridge, MA: Harvard University Press.

Coulter, H. (1979). *The social construction of mind.* London: Macmillan.

Dalin, P. (1972-1973). *Studies for innovation in education.* (Vols. 1- 4). Paris: Organisation for Economic Co-operation and Development. (Case studies of innovative schools and regions, arriving at results similar tothe insights from our Frankfurter Projekt.)

Diederich, H., & Lingelbach, K. Ch. (Eds.) (1977). *Erfahrungen mit schulischen Reformen.* Kronberg/Taunus.

Feyerabend, P. (1975). *On method: Outline of an anarchistic theory of knowledge.* London: NLB; extended German issue (19772): *Wider den Methodenzwang.* Frankfurt/M: Suhrkamp.

Furth, H. G. (1983). Young children and social knowledge. In T. Seiler & W. Wannenmacher (Eds.), *Concept development and the development of word meaning* (pp. 147-157). New York: Springer.

Garfinkel, H. (1967). Studies in ethnomethodology. *Englewood Cleffs, NJ: Prentice Hall.*

Goffman, E. *(1974). Frame analysis.* Cambridge, MA: Harvard University Press.

Good, T. L. & Brophy, J. E. (1978). *Looking in classrooms* (2nd ed.). New York: Harper & Row.

Hilgard, E. (1966). *Theories of Learning* (3rd ed.). New York: Appleton-Century-Crofts.

Hoetger, J. & Ahlbrand, W. P., Jr. (1969). The persistence of the recitation. *American Educational Research Journal, 6,* 145-167.

Jahnke, H. N. & Seeger, F. (1986). Piaget und Selz: Logische Strukturen versus Erkenntnismittel. In H. -G. Steiner, (Ed.), *Grundfragen der Entwicklung mathematischer Fahigkeiten* (pp.87-104). Koln: Aulis Verlag Deubner & CoKG.

Kagan, J. & Kogan, N. (1970). Individuality and cognitive perfoemance. In P. H. Mussen (Ed.), *Carmichael's Manual of Child Psychology* (3rd ed.) (Vol. 1, pp. 1273-1365). New York: Wiley & Sons.

Kounin, J. S. (1970). *Discipline and group management in classrooms.* New York: Holt, Rinehart & Winston.

Krummheuer, G. (1983a). Algebraische Termumformungen in der Sekundarstufe I—Abschlussbericht eines Forschungsprojektes. In H. Bauersfeld, M. Otte, and H. G. Steiner (Eds.), *Materialien und Studien, Band 31.* Bielefeld: Institut fur Didactick der Mathematik, Universitaet Bielefeld.

Krummheuer, G. (1983b). Das Arbeitsinterim im Mathematikunterricht. In H. Bauersfeld, H. Bussmann, G. Krummheuer, J. H. Lorenz, and J. Voigt (Eds.), *Lernen und Lehren von Mathematik* (pp. 57-106). Koln: Aulis Verlag Deubner & CoKG.

Kuhn, T. (1970). *The structure of scientific revolutions* (2nd ed.). Chicago: The University of Chicago Press.

Lorenzer, A. (1977). *Sprachspiel und Interaktionsformen.* Frannkfurt/M: Suhrkamp, stw 81.

Lyotard, J. F. (1979). *La condition postmoderne*. Paris: Edition Minuit. German translation (1986): *Das postmoderne Wissen*. Graz, Austria: Bohlau.

Mead, G. H. (1934). *Mind, self, and society*. Chicago: Chicago University Press.

Mehan, H. & Wood, H. (1975). *The reality of ethnomethodology*. New York: John Wiley & Sons.

Mehan, H. (1979). *Learning lessons*. Cambridge, MA: Harvard University Press.

Miller, M. (1986). *Kollektive Lernprozesse*. Frankfurt/M: Suhrkamp.

Ortega y Gasset, J. (1965). Die Aufgabe unserer Zeit. In *Gesammelte Werke II* (pp. 79-141). Stuttgart: DVA. (Original work published in Spanish in 1923.)

Otte, M. (1986, October). *Wege durch das Labyrinth*. Occasional Paper no. 82, Institut fur Didaktick der Mathematik, Bielefeld.

Radatz, H. (1976). *Individuum und Mathematikunterricht*. Hannover: H. Schroedel.

Schoenfeld, A. H. (1987). What's all the fuss about metacognition? IN A. H. Schoenfeld (Ed.), *Cognitive science and mathematics education*. Hillsdale, NJ: Lawrence Erlbaum Associates.

Schutz, A. & Luckmann, T. (1979). *Strukturen der Lebenswelt, Band 1*. Frankfurt/M: Suhrkamp, stw 284.

Seeger, F. (1987). *Activity, self-organization, and habitus: Theoretical concepts in mathematics education*. Unpublished manuscript. Bielefeld: Institut fur Didaktik der Mathematik, Universitaet Bielefeld.

Seiler, T. & Wannenmacher, W. (Eds.). (1983). *Concept development and the development of word meaning*. New York: Springer. German edition (1985): *Begriffs- und Wortbedeutungsentwicklung*. Berlin: Springer.

Seiler, T. B. (1984). Was ist eine "konzeptuell akzeptable Kognitionstheorie"? Anmerkungen zu den Ausfuhrungen von Theo Herrmann: Ober begriffliche Schwachen kognitivistischer Kognitionstheorien. *Sprache & Kognition, 2*, 87-1O1.

Steinbring, H. (1986). Mathematische Begriffe in didaktischen Situationen: Das Beispiel der Wahrscheinlichkeit. *Journal fur Mathematik-Didaktik, 6*, 85-118.

Voigt, J. (1984). *Interaktionsmuster und Routinen im Mathematikunterricht*. Weinheim: Beltz.

Voigt, J. (1985). Patterns and routines in classroom interaction. *Recherches en Didactique des Mathematiques, 6*(1), pp. 69-118.

von Glasersfeld, E. (1983). Learning as a constructive activity. In J. C. Bergeron & N. Herscovics (Eds.), *Proceedings of the Fifth annual meeting of the North American Chapter of the International Group for the Psychology of Mathematics Education* (Vol. 1, pp. 42- 69). Montreal: Universite de Montreal, Faculte de Sciences de l'Education.

Weinert, F. E. & Kluwe, R. H. (Eds.). (1984). *Metakognition, Motivation und Lernen*. Stuttgart: Kohlhammer.

Witkin, H. A., Dyk, R. B., Faterson, H. F., Goodenough, D. R., & Karp, S. A. (1962). *Psychological differentiation*. New York: Wiley & Sons.

Wittgenstein, L. (1974). *Tractatus logico-philosophicus* (Pears & McGuiness, Trans.). London: Routledge & Kegan Paul. German original in L. Wittgenstein and I. Schriften (1969). Frankfurt/M: Suhrkamp.

Wittgenstein, L. (1953). *Philosophical Investigations*. London: Basil Blackwell & Mott. German translation in L. Wittgenstein and I. Schriften (1969). Frankfurt/M: Suhrkamp.

Wittrock, M. C. (Ed.) (1986). *Handbook of research on teaching*, (3rd ed.). New York: Macmillan. (Overview for research in some of the discussed perspectives.)

Expertise in Instructional Lessons: An Example From Fractions

Gaea Leinhardt

University of Pittsburgh

Learning Research & Development Center

This paper describes one type of expertise in the teaching of elementary mathematics. The data base for this paper has been gleaned from practitioners who have been unusually successful in teaching students particular mathematical topics. The research that has generated this data has generally followed the paradigm of expert/novice contrasts, in which the skilled performance of experts is compared to the less skilled performance of novices, or in which differing types of expertise are examined. The students of the expert practitioners have emerged from their lessons *highly proficient* in performing the procedures and computations associated with a particular piece of mathematics and in learning the next level of mathematical material presented to them, and *moderately proficient* in explaining, generalizing, and coping with a variety of extension and higher-order tasks. It is also important to note that these students emerged from their experience liking math as a subject and feeling competent in it.

The teaching described in this paper takes place in a particular context, one which I argue implicitly is the acid test for evaluating or judging expertise in teaching. The context is that of real public school classrooms in which the teacher teaches all elementary subjects to between 25 and 30 students assigned to her or him. It is not the privileged university laboratory school, the classroom with a "visiting" teacher who teaches one section of mathematics per day, or the teacher of the gifted, and so forth, although much of value can and should be learned from these special environments. However, this paper is about what we have learned from very good practicing practitioners. The purpose of focusing on this particular context is to emphasize the complexity and embeddedness of the teaching task. Expert teachers in these environments are skillful both at managing the various portions of a lesson—the segments—and at explaining and teaching content. They keep lessons flowing and are aware of and in tune with what their students are learning. The teachers manage homework, seatwork, demonstrations, games, discovery projects, discussions, and drill with fluidity and consistency. Time is always treated as a valuable resource and is not squandered in getting set up and in making multiple unintended false starts. But that is not all. Expert teachers also *teach* very well. They give detailed, complete explanations and demonstrations, and provide rich mathematical experiences for their students.

Expert teachers use many complex cognitive skills. This paper will focus on only a few of them. Expert teachers weave together elegant lessons which are made up of many smaller *lesson segments*. These segments, in turn, depend on small, socially scripted pieces of behavior called *routines* that teachers teach, participate in, and utilize extensively. Expert teachers have a rich repertoire of instructional *scripts* which are updated and revised throughout their personal history of teaching. Teachers are flexible, precise, and parsimonious planners. That is, they plan what they need to but not what they already know and do automatically. Experts plan better than novices do in the sense of efficiency and in terms of the sharable trace that they operate from. From that more global plan—usually of a unit of material—they take an agenda for a lesson. The key elements of the *agenda* are available as mental notes the teacher has before teaching. The agenda serves not only to set up and coordinate the lesson segments but also to lay out the strategy for actually explaining the mathematical topic under consideration. The ensuing *explanations* are developed from a system of goals and actions that the teacher has for ensuring that the students understand the particular piece of mathematics.

The first part of this paper will deal with what we already know about lesson segments, routines, scripts, agendas, and explanations and with associated research questions. The discussion of agendas and explanations will include examples and data from a study involving novice and expert lessons on fractions. The second part of this paper will discuss questions for future research.

Lesson Segments and Routines

Lessons that are effective in teaching mathematics are not simply homogeneous blobs of teacher talk or student seat work. They are carefully crafted patterns consisting of short segments, each of which accomplishes a different goal and encompasses a different set of student and teacher actions. A lesson consists of a unique but familiar weaving together of these segments. The elegant work by Good, Grouws, and Ebmeier (1983) identified one particularly successful pattern of segments, but no doubt there are many.

Research on lesson segments and routines will be summarized briefly because both topics have rather thorough treatment in published papers (Leinhardt & Greeno, 1986; Leinhardt, Weidman, & Hammond, 1987). Routines support lessons by permitting both the teacher and the students to invoke a shared known set of behaviors for moving through each lesson segment. Predictable lesson segments play a similar role. That is, they represent blocks of known action from which meaning can easily be derived, like a familiar reference book in which the student can track the structural features of the lesson easily and locate the source for meaning (Leinhardt & Putnam, in press). In the overall structure of classroom lessons these two features, routines and lesson segments, are quite significant. Their importance is often overlooked because spontaneity, flexibility, and responsiveness are so highly valued in our culture, especially by educators. These two aspects of teaching (flexibility and predictability) are, of course, not at odds. Indeed, I would like to argue that it is only by having some part of the learning situation nailed down and "normalized" that one can buy the freedom necessary for fruitful intellectual exploration.

In studying the structures of many hundreds of lessons, Good, Grouws, and Ebmeier (1983) have generated a generic arrangement of lesson segments. It is the one best supported by the best data available. The lesson segments that we identified (Leinhardt & Greeno, 1986) are more variable and have a few other elements like games and problem solving episodes, but the segments are certainly compatible with what Good and his colleagues have found. Good and his colleagues see the use of specific patterns of segments in a slightly different way than do Leinhardt and Greeno, whose patterns of segments are driven by topics, not calendars. As Clandinin and Connelly (1985) have pointed out, teachers juggle many time frames—the year, holidays, the week, the day, and the topic. Good and his colleagues suggest a cycle of review and presentation that is structured by week and month. The teachers I have studied did, indeed, review on Monday. However, they also reviewed and tested at topical boundaries rather than in a fixed time sequence.

Lesson segments have an interesting secondary characteristic: they tend to have low temporal variance. This means that when a given segments occurs (such as homework review, public practice, or monitored practice), the time spent on it across lessons is somewhat stable. For example, the presentation segment in a lesson on a new topic tends to last 10-15 minutes and to occur in a block (Leinhardt, in press-c).

The merging of these two cyclical issues (calendar and topic) is a worthwhile item for future research because it represents such a teachable (to teachers) component. A necessary next step in documenting what constitutes excellent teaching is to build a taxonomy of teaching based on analyses of lesson patterns for different types of lessons, such as exploratory lessons or drill/performance lessons.

When the importance of lesson segments is ignored, several scenarios unfold. One situation occurs frequently with novice teachers. They fail to complete a coherent topic in a single class. Novices' time estimates are off and they often start the following class in the middle of the presentational segment from the day before, leaving the studentsconfused and nonresponsive. On the rare occasions when an expert is faced with the same situation, she or he would retain her or his basic lesson structure (review, public practice, presentation, etc.), recap the fragmented explanation in the review, and lead into a full presentation. When segments are erratic in pattern and length, students are distractible and nonresponsive. This *does not* mean that each lesson is a carbon of the next, only that there are similar *patterns* for similar types of lessons.

In the work on classroom routines (Leinhardt, et al., 1987; Putnam, 1985; Yinger, 1979), we (and others) have explored four basic types of routines: management, support, exchange, and learning. The first three have been carefully documented by Leinhardt, et al. (1987) in the context of watching expert teachers establish their classrooms during the first four days of school. Experts were observed to get most of their routines in place and understood within four hours of the start of the school year and to have them functional by the end of the first four days. Each teacher had a preferred set of teacher-student interactions or ways of "doing things," from passing out paper to responding to teacher queries. Given the speed with which very young children (age 7) master these routines, it would appear that they already expect that there is some way to do things in school, but do not know the specifics until told. Children have school schemas with empty slots that need to be filled with information on how to deal with these routine elements of school work. *Management routines* are largely people movers and placement actions, such as sitting, lining up, sharing, and so forth. When these routines are in place, the class seems to function smoothly. *Support routines* are the countless minute actions that support instruction, actions such as exchanging papers for correction, cycling groups of students to the board to publicly work on problems, and getting textbooks out of desks and opened to the correct page. *Exchange routines* facilitate communication. They are the rules of talking (e.g., "1/3 plus 1/3, Maggie?" vs. "1/3 plus 1/3" [pause, the cue for choral response]) and of topic control. These are the rules of communication that researchers seem to violate so frequently when working with children. An expert teacher has not only a well-established set of exchange routines but also a rich variety of ways to signal that a violation is about to occur: "Now we are going to do something a little different"; "The way we usually do this is . . . but now I want you to . . ." (When I read the protocols of researchers talking with students, I am always astounded at the distance in a linguistic sense between what the interviewer says and what most adults in conversation with a child would say.) These talking routines are flexible and changeable, but when a teacher changes them fair notice must be given. When flexible exchange routines are in place, the class flow seems natural and orderly.

Finally, we come to the routines of *learning*. These have not yet been studied systematically—and need to be. Learning routines are those that deal primarily with ensuring that students focus on content. For example, when a teacher says, "What did I just do?", the students respond, "You moved the ten over," or "You found the common denominator." They do not respond, "You picked up a piece of chalk," or "You looked towards the door." More subtly, students seem to understand the inherent rules of getting through an explanation and extracting the important parts of an example. So, for example, when a teacher presents four or five ways of rewriting and renaming numbers for subtraction with regrouping, the culmination of which is the cross-out and write-on-top procedure, the students do not seem too bothered by the fact that they had to go through so many other representations (expanded notation, felt strips, tens charts, etc.) in order to understand the new procedure, although on some occasions they get "stuck" on the expanded procedure and automatize it rather than moving on.

Leinhardt and Putnam (in press) have analyzed what students need in order to learn from a typical mathematics lesson. This analysis generated a model of the student learner and it allows us to carefully speculate on the effects on a student if a teacher were more or less competent in handling different parts of a lesson. For example, if a teacher flags the portion of an explanation that differentiates it from some other known aspect of the topic, the student will be more able to select and focus on the key points of the example. Building similar models of the student learner for lessons that are not didactic in form but take a more constructivist or inquiry-based approach would be useful. What does a student need to be able to do in order to learn from this style of teaching? We also need to learn what the lesson segments, routines, and structure are for lessons that are fundamentally inquiry-based. How is the totality orchestrated?

Lesson Scripts

A closely related concept to that of lesson segments is that of the curriculum script. This concept was first identified by Ralph Putnam (Putnam, 1985; 1987) and refers to the goals and sets of ordered actions for teaching a particular topic (Putnam & Leinhardt, 1986). The notion of a script emerged from research that attempted to show the ways in which teachers diagnosed and responded to student errors while tutoring. Putnam discovered that each teacher engaged in a limited set of specific teaching actions or moves. These actions included recycling through an explanation and constructing settings for student moves, but did not tend to include building a model of how the student was thinking about the problem. If teachers are supposed to be diagnosticians (Brown & Burton, 1978), why didn't they build these students-as-learners models as we expected they would? The answer is that they did build models, but in different ways than we had anticipated. They built them in terms of how the topic would be taught next time. Teachers seem to construct flags for themselves that signal material that will cause difficulty as it is being learned, and then they adjust their teaching of the topic in response to those flags or to past successes of what "worked." They seem to diagnose their teaching and its cycle rather than diagnosing the mental representation of a particular student. A major goal of teaching seems to be to move through a script, making only modest adjustments on line in response to unique student needs. Thus, a lesson given by an effective teacher who has been teaching for many years essentially contains layers of accumulated knowledge about the topic and how to teach it. This knowledge is flexible in a cumulative sense but is somewhat less flexible in the immediate or local sense.

Without these scripts novice teachers face two problems. First, they are frequently drawn off their focus to follow a particular student in ways that are not helpful for the rest of the class; and second, they fail to anticipate the crucial feature or dimensions of what is important or difficult about a topic.

Here again, the work for mathematics researchers is to build a rich taxonomy of lesson scripts that are known to be successful. The purpose is not to design a catechism of mathematical teaching but rather to begin a systematic line of inquiry on what works best (or even well) and under what circumstances. At the moment, a new teacher is armed with the sample lessons he or she worked out for the supervisor, one or two examples of lessons from the methods courses, and a personal memory of how a topic was taught to him or her. If a new teacher is unlucky, she or he has been told repeatedly *not* to use the textbook (see Ball & Feiman-Nemser, 1986). Having a well-researched and annotated set of lessons might be extremely helpful to both the novice teacher and the experienced, but not excellent, teacher who wishes to improve but for whom workshops have not proved helpful. One possibility would be to have a library of videotaped expert lessons.

Agendas

We will defer the important issues of more global planning to Yinger (1980, 1987; Yinger & Dillard, in press) and Clandinin and Connelly (1985) and focus here instead on agendas. In order to teach a lesson a teacher must have a plan. We have found rather consistently that the plans of expert teachers are rarely written down, or rather that the written lesson plan is so cryptic it is virtually impossible to analyze. However, the expert teacher does carry a mental plan that we have been able to access by simply asking for it. We call this the teacher's agenda. An expert teacher's agenda is brief but rich with information that she or he will use and modify while teaching. An agenda consists of goals and actions. It connects to the prior day's lesson by means of a goal structure. "We got this accomplished yesterday and now we need to do such and such." It contains a list of action segments appropriate for the stated topic of the lesson. The action segment statements refer to what the teacher will do and to what students will be learning. In addition, an agenda includes tests that let the teacher know what she or he needs to look for in order to determine whether or not to continue. The agendas that experts construct also contain a logical flow that is dictated by the subject matter itself, the learning needs of the students, or the interest level of the students.

The following empirical example demonstrates the differences between expert and novice agendas. In this study two novice teachers and three experts were interviewed. Before teaching a lesson on fractions, they were simply asked what they were going to do and what was going to happen in the lesson.

Their responses were analyzed in five different ways. First, we measured the quantity of the initial, unprompted verbal response by counting the number of typed lines. Second, we counted the number of references to *student* actions within this initial statement. Third, we looked for instances where the teachers indicated some planning for "tests," that is, checkpoints or midstream evaluations of student understanding or of lesson progress. Fourth, we examined the entire preinterview for items that could be classified as "instructional" actions, statements about both physical moves and content coverage. For example, one expert said, "I'm going to introduce the *unit on fractions*, the *terminology*, numerator, denominator. We're just going to work on *what part of a region is shaded* . . . we're going to use the *overhead*. We're going to *fold paper*." Within this statement, five instructional actions were counted. From a novice, we have the following: "I'm planning on *going over the homework* that I gave yesterday on fractions. I'm planning on going over it on the board." Within this statement, one instructional action was counted.

Finally, we examined the entire preinterview for the level and explicitness of its instructional logic and flow. We were trying to determine the degree to which any particular action preceded or followed another by some overarching rule. Rules that guided this logical flow could be content driven (e.g., moving the presentation from concrete to abstract or moving from the problem situation to solution) or child driven (e.g., extending the lesson because one group needs extra practice or controlling the lesson pace until the whole class reaches mastery). The presence of this logical flow was scored for each unique instance within an agenda. The importance of instructional logic is that it acts as a general posted constraint for the next probable move a teacher could make. This becomes crucial when a teacher's particular move has been poorly or inadequately specified in the plan.

Table 1 shows the contrast in results of the experts' and novices' agendas. The quantity of their first uninterrupted response was virtually the same, approximately 21 lines. The content, however, was very different. The experts described an average of almost 6 instructional moves while the novices' average was 4. The experts almost always referred to a decision point (test) in the lesson whereas the novices never did. The experts referred to the students' actions and behavior almost 3 times as often as the novices; and finally, the experts embedded a logical flow 9 times as frequently as the novices. This means that the experts were, in a sense, doubly armed to

Table 1

Means and Standard Deviations of Agenda Items in Expert (n=9)*
and Novice (n=5)** Preinterviews

Item	Experts		Novices		1 Tailed t TEST
	$\overline{X}$	SD	$\overline{X}$	SD	p <
Lines of Response	20.7	(8.5)	22.2	(11.1)	NS
Instructional Actions	5.8	(1.1)	3.8	(1.9)	.005
"Tests"	.4	(.5)	-0-	-0-	----
Student Actions	1.7	(1.4)	.6	(.9)	.025
Instructional Logic Elements	1.8	(1.5)	.2	(.40)	.005

*   9 interviews from 5 teachers

** 5 interviews from 2 teachers

teach. Not only did the experts have routines, patterns of lesson segments, and scripts that all worked more smoothly than those of novices (Leinhardt, et al., 1987; Leinhardt & Greeno, 1986; Putnam, 1985; Putnam & Leinhardt, 1986); but they also had a rich agenda at their fingertips (Leinhardt, in press-c).

To capture the flavor of the differences, let us look at excerpts from some novice and expert agendas. A novice is first:

> Okay, today, I'm planning on going over the homework that I gave
> yesterday on fractions. I'm planning on going—on
> going over it on the board, okay?
> [*First instructional action*]
> It'd probably be good for this class and
> I don't know how many I'm going to go
> over. There are 36 problems on the
> homework page, but I'm gonna see how it
> goes. If they're all getting them very
> quickly, then we'll move on. [*test*]
> [*Coding: 1 instructional move, 1 test,
> no student actions, no instructional logic*]

In the above agenda there seems to be little for the teacher to work from. It may well be that by writing out a detailed lesson plan (something required of novices), the novice did not build mental representations that were very strong or accessible. In essence, the novice "dumped" into the lesson plan and made no accessible long-term memory set. In contrast here is one expert's agenda:

> Okay, well, we're still working on fractions;
> looking at a complete set and reciting the
> fraction parts. Okay, tomorrow we'll be
> working on problem solving, word problems,
> have a pizza party. [*First instructional action*]
> And they have to identify
> what fraction has been taken out of the
> whole. [*Second instructional action*]
> We're still reinforcing the terms
> numerator, denominator. [*Third instructional action*]
> I have a muffin tin and I'll be putting,
> you know, like the muffin papers, something
> manipulative [*Fourth instructional action*];
> like today we cut apart and matched fractions
> with the illustrations. And just
> anything manipulative, so tomorrow they'll
> be working with the muffin tins and writing
> fractions [*Student action*], using the sets
> and the regions, [*Fifth instructional action*]
> [*Instructional logic*],
> checking their homework [*Student action*]
> [*Sixth instructional action*] from today.
> [*Coding: 6 instructional actions, 1 instructional
> logic, 2 student actions*]

Here is a second expert's agenda:

> We're going to begin adding fractions.
> [*First instructional action*]
> Yesterday we compared fractions, so we'll
> kind of review that after I introduce
> adding fractions.
> [*First instructional logic*]
> After we do some at the
> board, then we'll bring in what we did
> yesterday, comparison fractions . . .
> [*Second instructional action*]
> And I found out yesterday we can only do
> those if we draw objects.
> [*Second instructional logic*] It's hard for
> the children to see 3/4 is greater than or
> less than 1/4. So if they draw the object,
> [*Third instructional action*] [*Student action*]
> it's much easier. So, we'll be doing
> some of that today, just review. They
> don't compare fractions too much in fourth
> grade. But, I'll review it so the better

ones will know it pretty well. Then we'll
continue with adding fractions at the
board. [*Fourth instructional action*]
And then we'll go into the text book.
[*Fifth instructional action*]
[*Coding: 5 instructional actions, 2 instructional
logic, 1 student action, no test*]

The most remarkable thing about these three protocols is the total
absence of a usable plan in the statements of the novice. The novice had
taught a lesson on reducing fractions which had failed, had retaught the
same lesson, and had assigned homework. Her entire set of activities for
this day was slated for going over the homework. Even though she stated
she might "go on", she had no idea of what we would go on to. In contrast,
the first expert not only had a rather complete action list, she also knew
where she was within the broader topic of fractions. (Essentially, the students in that class were still working on the meaning of fractional parts.)
The second expert was a little further along in the unit and was teaching
adding fractions after having taught the first part of equivalence and the
comparison of the magnitude of fractions. This expert's plan reflects her
goal of possibly introducing a topic that might not be reached in the current
lesson in which she must first accomplish some repair work. Students' difficulty on the previous day's lesson on comparing fractions indicated to the
teacher that they needed to draw shapes and compare them. This expert's
plan for a repair lesson provides an interesting contrast to the novice's plan
for simply going over homework rather than finding a different way to teach
the lesson that had failed.

There is another aspect to the protocols worth noting. The experts displayed a deeper understanding of fractions as a teachable topic. Although
the experts would probably perform very similarly to the novice on a test of
fractions as a topic, the accessible and usable knowledge about how fractions
fit together and which piece is needed for which other piece was much richer
in the experts' minds. More research is needed on the role of teachers'
subject-matter knowledge and its organization in the teaching of mathematics.

Explanations

The heart of any teaching episode is the explanation of an idea or a
phenomenon. Explanations can be didactic and direct or discovery-based
and indirect. It is not our intent to argue here in favor of or against constructivist beliefs about the way one comes to know something. Rather, this
is a description of how teaching is done. Some teachers for some situations
design circumstances that lead a child to have insights about a particular
concept or problem; others tend to take a more direct teaching stance.

Regardless of which type of teaching one is describing, explanations given or the construction of an explanation are fundamental to the learning process.

In studying explanations, we have focused on some specific elementary mathematics topics, one of which is fractions. Fractions represent a formidable block of mathematical material for the older elementary and middle school child to learn. The scope and difficulty of the topic for students is complicated by the fact that for many elementary teachers their knowledge of fractions represents the upper bound of their mathematics knowledge. Furthermore, for many teachers, there is virtually no connection between the rather extensive set of procedures associated with fractions and any deep sense of mathematics.

Fractions are difficult for students to learn because they represent a major extension of the number system that the child has been dealing with, an extension that is introduced very briefly (compared to time spent on the natural counting numbers) and with little rationale. Fractions are also hard because the notational system is clumsy and arbitrary, having multiple referents or many meanings that can be attached to each part. For example, the denominator can refer to the whole of a common system (part/whole), to a part of a two-part system (part/part), or to a somewhat masked unit for an intensive quantity (miles per gallon, liters per second, etc.). Likewise, the line separating the two numbers can stand for a simple separation—an "is to"—or it can mean division. The motivating need for fractions, namely, that division is not closed for whole numbers, is not obvious, especially since most measurement problems can be solved by altering the fractional units to the smallest whole unit needed (e.g., 6-1/2 feet is 13 half-feet), which is sort of the way decimals work.

In addition to being difficult to justify as an entity and carrying a rather clumsy notational system, fractions present the child with a bewildering set of new operations and constraints (principles) on old operations. These constraints are confounded by the meaning or use of the fractional notation. For example, if one wishes to add 13/15 and 11/12 and one is referring to apple pies or to that gustatorially ubiquitous pizza, one converts the fractions to the lowest common denominator (60ths) and adds 52/60 and 55/60, getting 107/60, or 1 and 47/60. This means that out of 2 whole pizzas, 1 and 47/60 either remains or has been consumed. If, however, one is talking about the ratio of girls to boys in two classrooms *and* these are the real numbers of kids in two classes, then to find the ratio of girls to boys in both classes combined, one does what every fourth grader has always wanted to do, namely, add the two numerators (24) and the two denominators (27) for an answer 24/27. There is no *notational* cue for this bit of tomfoolery. It is purely referential and that suggests that the teacher who dares to cover such ground has a hard row to hoe. One could argue that referents are always a complicating factor in mathematics and that the selection of the "correct" model is always the problem. But for teachers and students, fractions are one major point at which this all comes out.

Having given a sense of what is hard about the topic we have chosen to study, we return to the discussion of explanations. Figure 1 shows a theoretical model of an expert elementary mathematics explanation. We designed this model while studying subtraction with regrouping and have adapted it to equivalent fractions here (Leinhardt, in press-a). At the top of Figure 1, there are three goals in hexagons: *clarifying* the concept or procedure, *learning* the concept or procedure, and *understanding* the concept or procedure. These goal states are achieved as both direct and indirect consequences of many action systems and subgoals, some of the more major of which are shown in this figure. The figure could be even more detailed than it already is; however, the major point is to demonstrate the key feature of an explanation.

The action of *explaining*, shown in the dominant rectangle immediately below the top three goal states, has as a consequence the complete or partial achievement of one or more of the top level goals. Explaining has a prerequisite goal state, namely, that the *components* to be used in the explanation are already known by the students. If a teacher or instructor attempts to explain something new by using an analogy or a new representation (Cuisinaire rods, pattern blocks, etc.) which itself must be learned, the explanation is unlikely to be of any major benefit to the hapless student who will likely lose sight of the principal objective.

Another prerequisite for the explanation is that the *subskills* to be used in the performance of the procedure are available—some might even argue that they must be available in automatic form. If these skills have already been taught, they can become available for use in the explanation process merely by asking for them, or by activating them, or by taking them out of cold storage, so to speak. The performance of one or more of these retrieving actions results in meeting the goal of having subskills available. In this version of the model, the subskills are factoring, knowing multiplication and division facts, knowing fractions terminology, and knowing the syntax of the particular representation being used.

The actions that support the goal of having subskills available can also be seen as prerequisites for the goal of having the necessary *numeric* and *concrete representations* described. Further, each demonstration must include the action of identifying the particular feature of the concept that makes it *unique* with respect to the solution. A good explanation will refer to this uniqueness and will explore salient moves and even warn of pitfalls inherent in the representation (Ohlsson, in press; Nesher, in press).

Another prerequisite for an explanation is to have the students realize what the *nature of the problem* is and, if appropriate, to locate the contradiction—the particular mathematical circumstance—that requires this

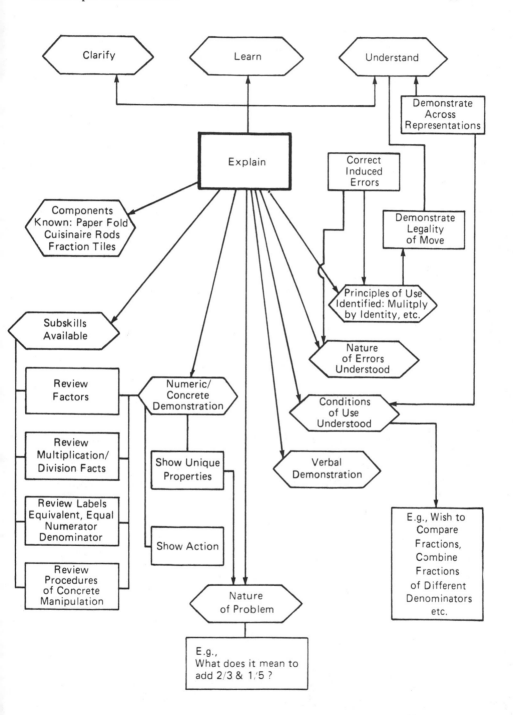

Fig. 1. Model of an explanation for fractions

solution. This can be done by actions that constrain the solution path so that students "bump into" the problem.

A corequisite for an explanation is having a *verbal demonstration* that goes through all of the key moves or concepts using language. Often, this is accomplished simultaneously and is paired with the numeric and concrete demonstrations.

A constraint on the explanation is that the *condition* and *circumstances* of use be identified. That is, the procedure needs to be recognized as one which does not get used all the time but is used only under some specific circumstances. Finally, the *principles* that permit the procedure to be used need to be identified and the *legality* of the procedure needs to be proven. The intentional production of *errors* is often a key move in an explanation; it is an action which can be used to accomplish the goal of understanding and identifying principles of use.

Not all explanations contain all of these elements but for each mathematics topic a competent teacher is able to access specific action schemas that support these goal states, to a greater or lesser degree. When we compared eight expert explanations and four novice explanations on the topic of equivalent fractions, we came up with the following results.

In the first row of Table 2 (Components Known), we can see that experts were using a commonly *known referent* 88% of the time while the novices

Table 2

Percent of Goals Present in Explanations

	Expert (n=8)*	Novice (n=4)**
Components Known	88	25
Subskills Available	100	50
Numeric and Concrete Demonstration	100	75
Verbal Demonstration Completed	100	0
Nature of Problem	38	0
Circumstances of Use	25	0
Principles of Use	38	0
Procedure Completed	100	0

*   8 lessons from 3 experts
** 4 lessons from 2 novices

were using it only 25% of the time. For example, when introducing the notions of equivalent fractions, one of our experts used folded paper and coloring sections. She had used the folded paper and labeling of parts on numerous occasions previously, so when she wanted to extend this concrete demonstration to show the equivalence of 1/2 and 2/4, the students had little trouble following her. In terms of *subskills being available* (row 2), the expert was able to call upon considerable fluidity in the knowledge of factors and multiplication strings. On the other hand, when one of our novice teachers introduced the same topic, she showed only two numerical representations and simply declared them to be equivalent; later in the explanation she used Cuisinaire rods, which the students had never seen before (components not known), and repeated the same action as she had with the numbers, saying that the two were equivalent. The only subskill available was knowledge of multiplication tables. Her students did not have the subskills of factoring or finding common factors, so a random-guess number strategy was used to identify an appropriate divisor.

As can be seen in the third and fourth rows of the Table 2, all of the experts completed and coordinated their *numerical and concrete demonstrations* and *completed their verbal descriptions* of concepts and procedures. The novices' explanations had no completed verbal demonstrations and the novices were less successful than the experts in completing the numeric or concrete demonstrations. For example, our novice who had attempted to demonstrate equivalent fractions with Cuisinaire rods never completed her demonstration as it became tangled. She started with a particular relationship, 1/2, and told the students how to find the equivalent fraction, 6/12, rather than letting the students find one on their own. Thus, the tough part of the procedure was masked. She failed to define what she meant by equivalent; nor did she give a complete verbal description of the procedure. Interestingly, the novice started out working with more imaginative tools than did the expert whom we observed teaching the identical lesson, but the novice had not planned time that would allow development of this as a referent system, nor was she able to connect the procedure to the concepts.

With respect to the *nature of the problem*, in an explanation of equivalent fractions we would expect some discussion or demonstration of equality, of the purpose of knowing that two things are equal, or of how things that look different in some respects can have the same value or can mean the same thing. Experts did this 38% of the time while novices did not do it at all. The same pattern (of experts attaining goals and novices not) held for the remainder of the characteristics.

We can see that, in terms of a model explanation, the novices do not do as well as the experts. However, to get the flavor of what a novice actually does and what an expert actually does, a small portion of one expert and one novice lesson will be described. The focus of the description for both teachers will be on lessons about equivalent fractions.

As background, it is useful to distinguish between the expert and novice with respect to how they saw the topic of equivalent fractions. We can build a sketch of their views from several sources: their agenda interviews, their in-class language, their commentary on their own videotapes, and their responses to a fraction interview.

The expert viewed the notion of fractional equivalence as absolutely central to all of the major procedural work that has to be accomplished in the teaching of fractions in fourth grade (concrete fractional referents such as number line, discrete object set representations, geometric regions, adding and subtracting with unlike denominators, converting improper fractions to mixed numbers, comparing fractions, etc.). She had a clear understanding of the power of equivalence to permit transformations and, in so doing, to permit simple algorithms to be applied to otherwise complex problems.

At its core, equivalence is a two-sided procedure. For this expert, the term *equivalence* encompassed the procedures for creating equivalent fractions, commonly understood to mean "raising" fractions, as well as procedures for reducing fractions. In both cases, her lessons rested on the use of fractional names for one (2/2, 3/3, 4/4, etc.) by which the given fractions were either multiplied or divided. The expert saw multiplying by a fractional name for one as simple and straightforward but saw division by a fractional name for one as more difficult. The critical distinction was that in multiplying, any fractional name for one is permitted, whereas in dividing, there is a specific and limited set of "appropriate" fractional names for one from which to choose. Finding the appropriate fractional name for one is a task in and of itself, and this search was explicitly taught.

The novice had a dramatically different sense of equivalence. Equivalence was the procedure used to raise a fraction to higher terms. It was a lesson sandwiched arbitrarily between adding and subtracting fractions with like denominators and reducing fractions to lower terms. It was a simple procedure, entailing no concepts, one which needed only a straightforward verbal description and a demonstration of manipulatives that were equivalent. The novice made no use of the terms or concepts of equivalence when teaching reducing or when she returned to adding and subtracting fractions with unlike denominators. The novice seemed surprised at each lesson when the students did not get the point and were unable to do the homework exercises. Her main solution was to reteach the identical lesson.

The expert began her lesson by reviewing the prior day's work. It was a paper-folding exercise that stressed language being attached to specific patterns of folds and the fact that two things that appear different can have the same value in terms of area or quantity. The expert spent a considerable amount of time working on the meaning of the equal sign and the term *equivalent*. In introducing the procedure for raising a fraction to an equivalent one, the expert drew on the students' knowledge of the property of one when used as a multiplier, namely, that it returned the same value. She then substituted different representations for one (2/2, 3/3, etc.). While going through this procedural explanation, she stressed again and again the equivalence and constancy of unit. In terms of controlling the range of the example, she explained the procedure for 1/2, with at least six iterations, then for 1/3, then for 1/4, and then went on to non-unit fractions.

The novice used the term *equivalence* twice in the course of her lesson. She never pointed out the relationship between multiplying by one and the procedure for raising a fraction. As a result, the students were quite lost as to which number was to be multiplied by what. Halfway through the lesson, she abruptly stopped the discussion of equivalent fractions and reviewed subtracting fractions with like denominators. Several days later when she was teaching reducing fractions, the novice interfered with a child's insight that equivalence and reducing were connected, and told him that they were different. The novice's main emphasis was on making sure the students did "the same thing" to the top and bottom number.

Both of these lessons focused on the procedural skill of making two fractions equivalent. But one of the lessons built upon a fairly strong linkage to basic mathematical principles, and it proceeded by justifying each procedural move in terms of these principles. The other lesson taught a very arbitrary set of steps.

Research that would enhance this work would include a careful analysis of when and how to teach procedural knowledge (like the work of Hiebert [Hiebert & Wearne, 1987] in decimals): how much conceptual knowledge should be embedded in a simple procedure and how much should precede or follow it. A second line of inquiry would focus on how deep a novice teacher's knowledge of fractions needs to be in order to explain at an adult level what the procedures are, how they could be different, and what mathematical principles are incorporated in the procedure to make it legitimate. In the particular case of equivalent fractions, I think it would be especially useful to do a close analysis of how different texts treat this topic.

Future Research

Having reviewed briefly what we know about the instructional processes of effective teachers and having pointed to research for each of these components, let me turn to some of the questions that remain to be resolved. Starting with the assumption that teaching is bound tightly both to the content being taught and to definitions of learning, the overriding implication is that the next NCTM conference on specific topics in mathematics education should not have one group dealing with teaching and separate groups dealing with content such as fractions, problem solving, or algebra. A major piece of research that needs to be done is on the fit between certain aspects of mathematics and the teaching of those topics. We should undertake this investigation with the assumption that different strategies and approaches will be required for handling different topics, ages, and students.

We have a fairly complete list of structural elements that go along with good teaching in terms of grouping arrangements and management techniques, but we are just beginning to document expert teaching of specific topics. The few good descriptions of teaching specific topics include: multiplication (Lampert, in press), subtraction (Leinhardt, in press-a), fractions (Leinhardt, in press-b), and an example of inquiry-based problem solving (Lampert, 1985). We need many more of these cases, preferably ones that include positive instances of more than one "style."

A second theme of research that is needed is to study the effects of teaching tough mathematical topics in dramatically different styles. Whether the research is accomplished by description or by experimental manipulation, we must move beyond the use of emotionally laden labels such as "inquiry" versus "rote" learning towards a better understanding of what can be gained by each approach. It seems silly and tautological to reason simply that because children learn under an inquiry approach without ever mastering math facts with speed, therefore, learning facts is unimportant or calculators will replace skill; or, on the other hand, that because procedural competence is present to a greater extent in other countries, we should assume that procedural knowledge is most important. Let us find out how much problem solving and inquiry we can get out of straight didactic teaching, how much computational fluidity and accuracy out of an inquiry approach, and then figure out when to use which kinds of approaches.

The next set of issues to consider in thinking about research on teaching is the complexity of lesson patterns. We know a great deal about didactic lesson patterns that work, but one of the dangers inherent in studying what expert teachers do is that we can never get beyond current practice. The danger in designing new approaches without considering current practice is that we will fail to understand the reality of the tasks that teachers face and how that reality interferes with the latest research vision. However, we need to try out configurations of lesson patterns that are different from ones that have already been shown to work. It would be interesting to see traditional mathematics taught with a combination of lesson styles, such as didactic, problem solving, inquiry, exploration, math laboratory, and game drill. For example, a teacher might start a given unit with two days of open-ended inquiry, using a range of materials and teams of students charged only with the task of coming up with as many ideas as they can think of about the materials. This might be followed by two days or so of small-group problem solving and focused attention (e.g., Lampert, 1985). Or, perhaps it could lead directly into a small set of didactic lessons in which some procedures and concepts were taught that would, in turn, be followed by a week-long problem-solving session. Also exciting, either as a weekly event or as a shorter daily visit, would be a math laboratory in which computer tutors, drill programs, and games were available and either assigned or freely explored. The idea is to build on what we know about effective lesson patterns while helping teachers to make their lesson patterns more complex, elaborate, and useful. In doing so, the teachers could enrich the students' concepts, concrete experiences, and extended problem-solving capabilities while not abandoning the computational aspects of arithmetic education that society seems to value.

REFERENCES

Ball, D. L., & Feiman-Nemser, S. (1986, October). *Using textbooks and teachers' guides: What beginning elementary teachers learn and what they need to know.* Paper presented at the Conference of The International Study Association on Teacher Thinking, Leuven, Belgium.

Brown, J. S., & Burton, R. R. (1978). Diagnostic models for procedural bugs in basic mathematical skills. *Cognitive Science, 2,* 155-192.

Clandinin, D. J., & Connelly, F. M. (1985). *Teachers' personal, practical knowledge: Calendars, cycles, habits and rhythms and the aesthetics of the classroom.* Unpublished manuscript, The University of Calgary, Calgary, Alberta, and the Ontario Institute for Studies in Education, University of Toronto, Toronto.

Good, T. L., Grouws, D. A., & Ebmeier, H. (1983). *Active mathematics teaching.* New York: Longman.

Hiebert, J., & Wearne, D. (1987). Procedures over concepts: The acquisition of decimal number knowledge. In J. Hiebert (Ed.), *Conceptual and procedural knowledge: The case of mathematics* (pp. 199-223). Hillsdale, NJ: Lawrence Erlbaum Associates.

Lampert, M. (1985). Mathematics learning in context: The Voyage of the Mimi. *Journal of Mathematical Behavior, 4,* 157-167.

Lampert, M. (In press). Knowing, doing, and teaching multiplication. *Cognition and Instruction.*

Leinhardt, G. (in press-a). Development of an expert explanation: An analysis of a sequence of subtraction lessons. *Cognition and Instruction.* Also to appear as a chapter in L.B. Resnick (Ed.), *Knowing and learning: Issues for a cognitive science of instruction.* Hillsdale, NJ: Lawrence Erlbaum Associates.

Leinhardt, G. (in press-b). Getting to know: Tracing students' mathematical knowledge from intuition to competence. *Educational Psychologist.*

Leinhardt, G. (in press-c). Math lessons: A contrast of novice and expert competence. *Journal for Research in Mathematics Education.*

Leinhardt, G., & Greeno, J. (1986). The cognitive skill of teaching. *Journal of Educational Psychology, 78*(2), 75-95.

Leinhardt, G., & Putnam, R. (in press). The skill of learning from classroom lessons. *American Educational Research Journal.*

Leinhardt, G., Weidman, C., & Hammond, K. (1987). Introduction and integration of classroom routines by expert teachers. *Curriculum Inquiry, 17*(2), 135-176.

Nesher, P. (in press). Microworlds in mathematical education: A pedagogical realism. In L. B. Resnick (Ed.), *Knowing and learning: Issues for a cognitive science of instruction.* Hillsdale, NJ: Lawrence Erlbaum Associates.

Ohlsson, S. (in press). Sense and reference in the design of interactive illustrations for rational numbers. In R. Lawler & M. Yazdani (Eds.), *Artificial intelligence and education.* Norwood, NJ: Ablex.

Putnam, R. (1985). Teacher thoughts and actions in live and simulated tutoring of addition. (Doctoral dissertation, Stanford University, 1985.) *Dissertation Abstracts International, 46,* 933A-934A.

Putnam, R. (1987). Structuring and adjusting content for students: A study of live and simulated tutoring of addition. *American Educational Research Journal, 24*(1), 13-28.

Putnam, R., & Leinhardt, G. (1986, April). *Curriculum scripts and the adjustment of content in mathematics lessons.* Paper presented at the annual meeting of the American Educational Research Association, San Francisco.

Yinger, R. J. (1979). Routines in teacher planning. *Theory into Practice, 18*, 163-169.

Yinger, R. J. (1980). A study of teacher planning. *Elementary School Journal, 80*, 107-127.

Yinger, R. J., & Dillard, J. (in press). Thought and action in teaching. In P. L. Peterson & H. J. Walberg (Eds.), *Research on teaching: Concepts, findings and implications*, (2nd ed.) Berkeley, CA: McCutchan Publishing Corporation.

Yinger, R. J. (1987, April). *By the seat of your pants: An inquiry into improvisation and teaching*. Paper presented at the annual meeting of the American Education Research Association, Washington, DC.

The research reported herein was supported in part by the Learning Research & Development Center's Center for the Study of Learning, which is supported in part by funds from OERI, United States Department of Education. The opinions expressed do not necessarily reflect the position or policy of OERI, and no official endorsement should be inferred.

The author wishes to acknowledge the helpful comments and research support of Joyce Fienberg and Gwendolyn Hilger.

Implications of Research on Pedagogical Expertise and Experience for Mathematics Teaching

David C. Berliner, Pamela Stein
Donna Sabers, Pamela Brown Clarridge,
Katherine Cushing, and Stefinee Pinnegar
University of Arizona

In the field of psychology during the last few years, experts have received a great deal of attention. The publication of de Groot's (1965) book *Thought and Choice in Chess* stimulated many researchers to ask questions about domain-specific knowledge: How is it organized, processed, and remembered; how is it used in perception, problem solving, and transfer; and so forth. As these studies were completed, it soon became of interest to ask whether there were common features among the experts' ways of thinking across different subject areas or fields. In a short time experts and novices in fields as diverse as radiology, bridge, baseball, physics, mathematics, knowledge of dinosaurs, and in other knowledge and skill domains were compared. Some common characteristic of experts were soon revealed. Across numerous domains it was found that:

1. Experts often make inferences about the objects and events in their purview; novices usually hold more literal views of those objects and events.

2. Experts often classify problems to be solved at a relatively high level; novices usually classify problems by surface characteristics.

3. Experts, as compared to novices, have faster and more accurate pattern recognition capabilities.

4. Experts are slower than novices in starting to solve a problem; they seem to take longer to examine the problem and to build a problem representation.

5. Experts build different problem representations than do novices.

6. Experts, as compared to novices, show greater self-regulatory or metacognitive capabilities.

7. Expertise is built up slowly, over considerable amounts of time, with considerable amounts of practice.

Other common characteristics of experts in different domains have also been reported (Chi, Glaser, & Farr, in press; Glaser, in press; Berliner, 1986a). The inescapable conclusion from all this relatively recent work is that experts in diverse fields share common characteristics. But clearly missing from this initial flowering of research in an interesting area of contemporary cognitive psychology have been studies of expert teachers.

STUDYING EXPERT TEACHERS

There are many reasons for wanting to study expert teachers. First, and of great psychological interest, is the inquiry into whether expert teachers resemble experts in other fields. Because of its uniqueness in requiring decision making and problem solving in a public, complex, and dynamic environment for many hours every day, teaching has been said to be among the most difficult of professional jobs. Despite the complexity of teaching, anecdotes and logic inform us that it is likely that some teachers have managed to find methods to organize their classrooms and to convey information in ways that are better than those used by other teachers. It would be informative to inquire if the domain-specific knowledge of these expert teachers is used in ways that are similar to the ways chess players and radiologists use their domain-specific knowledge.

A second reason to study experts is to get access to the cognitions accompanying the behaviors that the field of research on teaching has revealed as effective. Process-product research, particularly in mathematics instruction, has confirmed the usefulness of many teaching behaviors. Nevertheless, we do not yet know much about how teachers think about those teaching behaviors that research has taught us are important (Berliner, 1986b). For example, both Rosenshine and Stevens (1986) and Good, Grouws, and Ebmeier (1983) have recommended the opening homework review for certain kinds of mathematics lessons. But what are the bases for certain kinds of decisions made by teachers, initiating an opening homework review in mathematics on some days and not on others? What cues in the environment inform teachers to keep reviews going or to stop them early? How do teachers obtain information to distinguish between students who are simply unprepared and those who are having trouble with the concepts that were to be practiced? What are the routines—the scripted, virtually automatic pieces of action understood well by students—that occur within an opening homework review in a mathematics class, and how are those routines used to ensure that the review is done quickly and efficiently? Leinhardt and Greeno (1986), for example, in studying elementary school mathematics lessons, compared an expert teacher's opening homework review with that of a novice. The expert teacher was found to be quite brief, taking about one-third less time than the novice. This expert was able to pick up information about attendance and about who had or had not done the homework and was also able to identify who was going to need help later in the lesson. She elicited mostly correct answers throughout the activity while also managing to get all the homework corrected. Moreover, she did so at a brisk pace and never lost any control of the lesson. She also had routines for recording attendance, for handling choral responses during the homework checks, and for using hand-raising to get attention. This expert also used clear signals to start and finish the lesson segments. Interviews with this expert teacher revealed how the goals of the lesson, the time constraints, and the overall curriculum itself were blended and became a guiding element for the metacognitive activity she used to interpret her teaching performance. In contrast, when the novice was enacting an opening homework review as part of a mathematics lesson she was not able to get a fix on who had and had not done the homework, she had problems with taking attendance, and she asked ambiguous questions that led her to misunderstand the difficulty of the homework. At one point the novice lost control of the pace. She never did learn which students were going to have more difficulty later on in the lesson. Of importance, as well, was the novice's demonstrated lack of familiarity with well-practiced routines. She seemed not to have habitual ways to act; students, therefore, were unsure of their roles in the class.

The traditional forms of observational, correlational, and experimental work caused us to focus our attention on the opening homework review and provided us with some assurance about its utility. The opening homework review is a kind of "macro" teaching variable. But it is only with a microanalysis of an opening homework review task and through interviews with expert and nonexpert performers of the tasks that we get fresh insights about the criteria that might be used for judging between better and worse enactments of the task itself (see also Brooks & Hawke, 1985).

The expert described in Leinhardt and Greeno's study provides us with another reason for studying expertise in teaching: Experts, more than most others, can provide exemplary performances. In the case just described, the exemplary performance was of an opening homework review. In another case (Leinhardt, 1985), also in mathematics, it was of an expert teacher teaching subtraction using regrouping. It is an unfortunate fact that we have no extensive case literature in education, such as we have in medicine, law, or business. An exemplary performance—virtuosity in the teaching of something to someone by an acknowledged expert—could be a great source of case study information for novices. Shulman (1986; 1987), who has described the kinds of cases that could be used in teacher education programs, notes that cases of chess go back hundreds of years and are available for study by novices and by others. There are also case books in chess that specialize in temporal aspects of the game such as openings, mid-game moves, and end-game moves. In addition, there are casebooks of great tournaments by expert players (the international championships, for example, with their move-by-move analysis). It is possible that by increasing our knowledge of experts in pedagogy we will be better able to identify some cases of exemplary performance and to understand, as well, why they are exemplary. Videotapes or transcripts of exemplary performances could be made available to novices and other professionals interested in such things as lesson openings, lesson closings, or teaching tough concepts (e.g., the difference between fraction, decimal, and percentage equivalents). Exemplary lessons or units on factoring, long division, or permutations all could become part of the training materials for novice teachers. Such cases might also present new ways to teach a lesson on electricity that actually change the naive intuitive beliefs about electricity that students have which, unfortunately, often remain intact even after instruction.

A third reason for studying experts is to influence state and local school district policies regarding master teachers. There is currently much interest in "career ladders" in which new roles are being sought for experienced master teachers so that they need not always be in the classroom instructing students. Such special teachers could serve in supervision of novices, in curriculum development, in assisting with the evaluation of peers, in providing coaching of teachers having difficulty, and so forth. It is anticipated that in each district a small number of teachers could take on these special senior roles and be compensated at much higher rates than would the ordinary teachers of the district. The U.S. public is in favor of these ideas. It recognizes that the teaching profession is relatively flat with no chance for advancement except to administration. Most people believe that teachers who are experts ought to have a way to rise in the hierarchy and to be remunerated accordingly. The public also seems to believe that mere number of years on the job should not be enough to qualify one for the honorific position of master teacher. If the public is willing to put out money for these new career positions it will ultimately want some assurance from the educational research community that expert or master teachers do indeed exist and that they have characteristics that would distinguish them in some way from more ordinary teachers.

A fourth reason for studying expert teachers is to influence current policy in certain states. The United States, after many years of a surplus, is now facing a shortage of teachers. It is estimated that one million new teachers will be needed in the work force during the next decade. The shortage of teachers in mathematics is already known and is expected to become particularly acute. The university teacher-education programs cannot possibly provide the number of people needed to fill those jobs. Thus in some states, to deal with this problem, teaching certificates are provided to people with mathematics subject-matter knowledge but no teaching experience or training. Such action ignores the complexity of teaching mathematics in a modern public school. Therefore, in the research program described below, we often compared the performances of expert and novice teachers with a third group we called *postulants*. The postulants were personnel from business and industry with mathematics and science skills and a desire to change careers to teaching in a public school. They wanted entry into the profession of teaching but did not want to take any teacher education course work. This kind of comparative research, we hope, might yield information that would inform those who make state policies on providing teacher certification to untrained individuals.

Some aspects of this ongoing research program are described next. Selective findings from this program are also presented.

METHODOLOGY AND SELECTIVE FINDINGS FROM SOME STUDIES OF EXPERTISE IN TEACHING

In education it is difficult to distinguish between experience and expertise. No outside criteria are available, as in chess tournaments, that declare that someone is a grand master. Arguably, achievement of students, perhaps as measured with standardized tests, might be used as a criterion to distinguish expert from less expert teachers in self-contained elementary school classrooms. But in high schools, where much of our focus has been, even this imperfect criterion measure is difficult to use since students are affected by half a dozen different teachers each semester. A college bound student may have courses in computers, mathematics, biology, and chemistry in one semester. A non-college bound student may have business mathematics, general science, and a course with an electronics laboratory. In each case there is the possibility that some of the curriculum in the different courses may overlap. Thus a student's standardized test performance may not always be attributable to the effects of a single teacher. In obtaining experts to study, therefore, we often used informal nominations and experience as initial criteria for determining expertise. Classroom observations were used as well. But in all but one of the studies described below, experience and expertise are inextricably linked and difficult to untangle. This is not unique to studies of expertise in education. Expert mathematicians, physicists, or radiologists who are studied are often individuals who are nominated by peers as experts and who have more experience than some comparison group often designated as novices. Thus in many studies with no clear criterion to determine expertise we should be aware that the expert group is not as clearly defined as in studies of expert chess or bridge players. This problem, no doubt, works against finding differences between experts and others, thereby making the differences that are found all the more impressive.

A Study of Expert Teachers of the Gifted

The first of our inquiries was a small study done by Gail Hanninen (1985) comparing three groups of teachers of the gifted. The first group were experienced/expert teachers obtained by means of nominations from university professors and supervisors of teachers of the gifted. The second group was made up of experienced teachers who were studying to become certified as specialists in the teaching of the gifted. The third group were novices who were just finishing coursework for certification as regular classroom teachers and had no special knowledge of teaching the gifted. All groups consisted of 5 teachers. The methodology used in this study consisted of 5 scenarios or cases. Subjects had to read and then make written recommendations for cases such as the following example:

> Mark is a very active 8-year-old who is in the third grade at John's Elementary School. Academically, Mark particularly likes mathematics, reading stories about adventure, humor, and history. In addition, working on the computer and doing science experiments has a special intrigue for Mark. An accurate assessment of Mark's academic performance is effected by a bilateral severe to profound hearing loss. However, on the Performance Scale of the Wechsler Intelligence Scale for Children, he scored in the superior range. This was also supported by performance on the Arthur adaptation of the Leiter Scale.
>
> Mark's inquisitiveness helps in the cultivation of his interests in drawing pictures and designing buildings and models. Such curiosity is not only observed at school, but also at home where he pursues his interests in tennis, cooking, and photography.

There were two findings from this study that are of particular interest. First of all, some of the experienced/expert teachers of the gifted used a higher-order system of categorization to analyze the problem they faced. For example, in one of the novice's protocols about Mark, we find this opening sentence: "Mark seems like a very talented individual with many diverse interests." Another novice began her recommendations with: "Mark should be encouraged by his teacher to continue his science experiments and work on the computer." An experienced teacher who was still a novice in the area of gifted education began her recommendations with: "He should be able to pursue his interests in greater depth." The beginnings of the written recommendations from the novices and from the experienced teachers who were novices in the area of gifted education were quite banal. Only surface characteristics of the problems were used to discuss ways to help Mark. This is in contrast to the remarks of an experienced/expert teacher who began her recommendations with this opening statement: "Mark's needs can be broken into three broad areas: academic enrichment, emotional adjustment, and training to cope with his handicap." Some commonalities among experts were noted earlier: that experts seem to represent problems differently than do novices, and that they seem to use higher-order systems of categorization to analyze the problems they face. In this study some of the experienced/expert teachers showed those qualities. When they broke up Mark's educational needs into three categories and then described actions relevant to each of those categories these experienced/expert teachers of the gifted resembled the expert physics, mathematics, or medical problem solvers described in other studies.

A second finding of interest from this study was that the mean time for a novice to read through a scenario and begin writing about ways to help deal with a particular child was 2.6 minutes; the mean time for experienced teachers, who had no background in gifted education, was 3.0 minutes; and the mean time for experienced/expert teachers of the gifted was 9.8 minutes. That is, from the start of reading their problem through the start of presenting their solutions, it took the experienced/expert teachers 3 or 4 times as long as the other two groups, a pattern similar to that in other fields of study.

If we were to extrapolate from this study, we would expect experienced/expert mathematics teachers also to have different, and probably richer, problem representations than, say, novice mathematics teachers. Perhaps we might also expect the experienced/expert teachers to be thoughtful about students' needs, reflecting their years of experience as teachers. Their thoughts about students' needs, in comparison to novices would be expected to be richer, deeper, less banal, and so forth. Some support for this comes from another of our studies, called the prediction study, described next.

The Prediction Study

In this and in other studies in this series we compared experts, novices, and postulants in mathematics and science teaching at the secondary level. The expert teachers were drawn from a small pool of teachers in the Tucson area who were nominated as expert by their principals and judged to be exceptional teachers by three independent, experienced observers who had visited their classrooms. From 55 original nominees, a pool of 17 were designated as exceptional teachers to be studied. All had 5 or more years of mathematics teaching experience. Novices in this study were first-year science and mathematics teachers who had been rated as excellent student teachers. Postulants, as noted above, were scientists and engineers from business who were interested in teaching careers but had no formal training or experience as teachers.

In this study 10 experts (5 mathematics, 5 science), 6 novices (3 mathematics, 3 science) and 6 postulants (3 assigned to mathematics, 3 assigned to science) were required to make predictions about student knowledge in science and mathematics (Stein, Clarridge, & Berliner, under review). The instrumentation used consisted of a set of mathematics and a set of science items taken from the 1982 National Assessment of Educational Progress (NAEP) tests. The items were representative of those given to a national sample of 13- and 17-year-olds during 1982. The first five items were on separate index cards and were examined one at a time by the subjects. The subjects were asked to think aloud while they predicted the percentage of 13-year-old and 17-year-old students who correctly responded to the items. The subjects were also asked to think aloud while they guessed which distractors would be most likely and least likely to be chosen by the students taking this national exam. After the five think-aloud items were completed, the experts, novices, and postulants were prompted to reflect upon and discuss patterns in the ways in which they approached the task. Finally, these subjects continued making predictions for 15 more mathematics and science items, but they did not think aloud during that time.

One of the interesting findings from this study was that there were no discernible differences between expert, novice, and postulant teachers in their predictions of the percentage of 13- and 17-year-olds who might get the items correct. However, there did appear to be a slight difference between the science and mathematics teachers, with the science teachers being just a bit more accurate in their predictions than the mathematics teachers.

In analysis of the protocols obtained while the subjects were thinking aloud it become apparent that there were a number of differences between the groups. Compared to novices, about twice as many postulants and experts named or labeled items. Among the mathematics subjects, all experts and postulants employed this strategy. Only one novice mathematics teacher labeled an item. The experts' labels were much more detailed and specific than were the postulants. For example, for one mathematics item a postulant used the label "a decimal problem;" an expert labeled the same problem "a non-repeating, non-terminating decimal." In science there was an item called "an electricity problem" by the postulant about which an expert said, "It's a series aiding circuit problem." To that particular science problem not a single novice provided a label. The content expertise of our postulants and of our experts apparently allowed them to name the problem in such a way that they could begin to access the domain-specific knowledge needed to interpret that type of problem. Novices seemed not to do that. In fact, on a number of dimensions, the experts and postulants in this study resembled each other whereas the novices did not approach this task as sensibly.

We also found that the experts were able to do something that the postulants were not able to do: They could do a task analysis of the problem in a way that was quite sophisticated. More than the other two groups, they analyzed the demands of the task represented in the item. This may be a category of considerable importance. Task analysis was coded when the subjects verbalized something about the reasons for an item's difficulty or when they traced out the various steps or competencies that a student would need to answer an item correctly. Eight of the 10 experts analyzed the task demands of the items, which they did for from one to four of the five items for which they had to think aloud. This may be compared to the novices and the postulants, of whom only 3 out of 6 in each group engaged in task analysis of the items, and which they only did for one of the five items.

The task analysis for items were also more embellished or delineated when given by experts. For example, to one mathematics item an expert said:

Expert: The equal to or greater than symbol would be more difficult than just solving a simple equation. . . . Some would be able to do it by trial and error, by actually putting numbers in, but this wasn't multiple choice. . . .

Another expert said:

Expert: A student would definitely need to be able to translate this from the English words into symbols to be able to solve this problem.

Novices did not provide such sophisticated analyses. For example:

Novice: . . . the problem itself really isn't hard at all. It's the wording that makes it hard.

A postulant, in a similar manner stated:

Postulant: This is a somewhat harder problem. It's a word problem, which makes it even harder.

There are a number of other ways in which the experts, novices, and postulants differed. For example, they differed in their inferences about the student cognitions used in answering an item. Experts seemed to have a fund of knowledge about the way students thought and how those thoughts interacted with the content of the specific mathematics or science items. In addition, the experts seemed able to think through the misalgorithms that students might apply. The experts had more experience in dealing with student errors and therefore knew what types of errors students might make. Novices rarely discussed the issue of misalgorithms that students might apply to solve a problem.

The dearth of information from our novices would have been disconcerting had we not been aware of findings about the performance of those at an intermediate stage in the development of expertise. Interestingly, experts and those with the least experience seem to perform better on some tasks than those who are at an intermediate stage of developing expertise (Feltovich & Patel, 1984; Lesgold, 1984; Strauss & Stavy, 1982). In this study the novices, with minimal experience, may be considered intermediate in development as compared to the postulants (inexperienced) and the experts (very experienced). Glaser (in press) also refers to this phenomenon and explains further, "In the course of acquiring expertise, plateaus and non-monotonicities of development are observed which appear to indicate shifts in understanding and stabilizations of automaticity" (p. 11). Evidence of this phenomenon in our novice group has bearing on how teachers' informational systems are developed. Perhaps too much is expected of student teachers and first-year teachers, or perhaps they are judged too harshly for lack of skills when, in fact, they are in the process of developing, reorganizing, and consolidating massive amounts of new and complex information.

From the prediction study and the study of experienced/expert teachers of the gifted, then, we conclude that experienced/expert mathematics teachers, as compared to novices, are more likely to have richer representations for classifying mathematics problems and the thought processes that students use to solve (or fail to solve) such problems. The postulants, who in some of our studies seemed less able than novices, showed behavior more like the experienced/expert teachers in the prediction task. The implications of this research for alternative certification policies are, unexpectedly, on the positive side. And the implications of this research for the coaching of new teachers (a crucial time in the development of expertise) seem obvious. It is also obvious that research on the developmental processes in learning to teach mathematics is needed. The route from novice to experienced/expert teacher is virtually unknown in mathematics and in every other subject-matter content area.

Other studies in our research program focused on how teachers acquire information from classrooms. Two such studies have been completed and are described next. Both studies focused on visual information processing.

The Slide Task

This study examined how the experienced/expert, novice, and postulant teachers perceived visual information about classrooms (Carter, Cushing, Sabers, Stein, & Berliner, in press). Photographic slides were made of two 55-minute secondary class sessions: a science laboratory on chemical reactions, and a lesson on simple geometric proofs in a mathematics class. The slides were used as the basis for creating four tasks. Each task required subjects to view a series of slides and to respond to a structured interview both orally and in writing. The tasks were sequential, taking about an hour. In Task 1 subjects were shown three slides, each for approximately one second and then asked to write down what was seen. This was called the "Quick Look Task." It was designed to assess the immediate perception of the members of each group to rapidly presented visual classroom stimuli. In Task 2 the subjects were shown a different set of slides. Each slide was shown for three seconds, and then participants were asked to write down everything they noticed. For this task, each slide was shown for a second and a third time. Subjects were asked to record any additional information they noticed with each viewing. This procedure was repeated for each of the five slides that comprised this task. We called this the "Look Again Task." In Task 3, called the "Tell Me a Story Task," subjects were asked to view a sequence of approximately 50 slides, each shown for five seconds. Slides for this task were selected to present visually rich classroom information. Slides were arranged sequentially for viewing so that subjects could view a lesson segment from beginning to end. After the entire sequence of slides had been viewed subjects were asked to reconstruct the sequence of events portrayed in these slides. Subjects were also questioned about specific events and specific students who appeared in the slides. In Task 4 subjects viewed the same slides as in Task 3, but they had the power to stop the sequence at any time to discuss particular slides or combinations of slides that they believed to be salient for purposes of classroom management and instruction. This was called the "Stop and Talk to Me About Management and Instruction Task."

We noted earlier that experts make inferences about the objects and events in their purview, whereas novices hold more literal views of the objects and events around them. In this study we found evidence from our experts that they possess these same kinds of information-processing skills as do the experts in other fields. For example, the responses of the postulants and novices to the brief exposure to a slide presented in Task 1 were clearly descriptive, and usually quite accurate.

Postulant: A blond haired boy at the table, looking at papers. Girl to his left reaching in front of him for something.

Novice: [It's] a classroom. Student with back to camera working at a table.

Novice: A room full of students sitting at tables.

In contrast to these literal descriptions, typical of novices and postulants, some of our expert teachers often responded with inferences about what they saw.

Expert: It's a hands-on activity of some type. Group work with a male and female of maybe later junior high school age.

Expert: It's a group of students maybe doing small group discussion on a project as the seats are not in rows.

For experts, the information often deemed important was information that had instructional significance, such as the age of the students or the teaching/ learning activity in which they were engaged.

As we also noted earlier, experts have extraordinarily fast and accurate pattern recognition capabilities. The recognition of patterns reduces the cognitive processing load for a person. Sense is instanteously made of a field, such as a chessboard. Quick pattern recognition allows an expert chess player to spot areas of the board where difficulties might occur. Novices are not as good as recognizing such patterns, and when they do note them they are less likely to make proper inferences about the situation.

In Task 2, the "Look Again Task," we learned something related to this issue. This task called for updating of information as a slide was viewed three times. One expert in science, after the second viewing of a slide, said:

Expert: It's not necessarily a lab class. There just seemed to be more writing activity. There were people filling out forms. It could have been the end of a lab class after they starting putting the equipment away . . ."

And then after the third viewing of the slide the expert said:

Expert: Yeah, there was . . . very little equipment out and it almost appeared to be towards the end of the hour. The books appeared to be closed. Almost looked like it was a cleanup type of situation.

Novices did not usually perceive the same cues in the classroom and could not, therefore, make the inferences which guided the expert's understanding of the classroom. The expert, by the way, was absolutely correct. It was a cleanup kind of activity. We regard the reading of a classroom, like the reading of a chessboard, to be in part a pattern recognition phenomenon based on hundreds and thousands of hours of experience.

In Task 4, the "Stop and Talk to Me about Management and Instruction Task," we learned that experts seemed to agree among themselves about salient characteristics of classrooms more so than did novices or postulants. In this task the subjects went through a series of about 50 slides at their own paces. They could stop and comment on any of the slides that they wanted to. The novices and the postulants seemed to show no particular pattern in what they stopped to comment on, and often showed some contradictions. That is, one novice might say, "everything looks fine, they're all paying attention," and another novice might say, "it looks like they're starting to go off task, they're starting to drift." A pattern was noted among the experts that was quite different. For example, here are some comments about a slide at the beginning of the sequence and a slide at the end of the sequence.

Expert: [Slide 5] It's a good shot of both people being involved and something happening.

Expert: [Slide 5] Everybody seems to be interested in what they're doing at the lab stations.

Expert: [Slide 5] Everybody working. A positive environment.

Expert: [Slide 51] More students with their books closed, their purses on their desks, hands folded, ready to go.

Expert: [Slide 51] Must be the end of class and everybody is getting ready for the bell to ring.

The experts, more often than the subjects in the other groups, found the same slides worth commenting about, and had the same kinds of comments to make.

There were two additional findings that may be of interest to the mathematics education community. First was the focus of the experts on the notion of work in many of their comments: "students *working* at the blackboard," "students *working* independently," "teacher looking over a person *working* in the lab," and so forth. Work appeared to be a salient organizing concept for the experts as they viewed these slides.

In the main, postulants' descriptions were characterized by greater detail about the more static features of the classroom environment. Students were one aspect of the environment, but, in contrast to experts' descriptions, the students and their work-related actions were not the prominent features of the postulants' descriptions. In fact, students and their work involvement appeared no more salient to postulants than did descriptions of the physical surroundings of the classroom, the equipment, the windows and desks, and students' locations in relationship to chalkboards, areas of the room, or other students. Apparently, all of the visual stimuli presented had equal information value.

Second, experts also seemed to use their knowledge of classrooms to make assumptions about what was happening in the slides they saw. Notice in the following excerpts the explicit use of the word *assume*, by three different experts.

Expert: Students were not seated in the traditional type of seating arrangement— one that would normally be used for a lecture- type style of teaching. So from the seating arrangement I assumed that they must have been involved in some activity other than a traditional type of a lecture.

Expert: . . . there aren't a whole lot of humorous math problems so I assumed a couple of the students must have been talking—from their facial expressions— about something other than the assignment.

Expert: I assumed it was the teacher's desk because [it] was faced a different way from where the students' desks were faced and because of where it was placed in relationship to the chalkboard.

In many other comments made by experts the term *assume* could be substituted for the actual words used, as illustrated below:

Expert: It looked like not all students were focused on the same thing; some were faced forward, some were faced back, some were looking off to the side. So it didn't seem like they were all [so I assumed they were not] having any sort of group activity.

Assuming, that is, *hypothesis making*, was more prevalent among the experts than it was among novices or postulants.

Extrapolating from this study we might expect, therefore, experienced/ expert teachers of mathematics to be more likely than novices or postulants to focus on events that have instructional significance; to "read" the patterns of classrooms quicker and more accurately; to agree among themselves about what is and is not going on; to be more work oriented in their views of classes; and to make more assumptions and hypotheses about classroom phenomena. These hypotheses need to be carefully examined and, certainly, more support for them is needed. Nevertheless, even now, these hypotheses are testimony to the probable uniqueness of experienced/expert teachers.

The Simultaneity Task

The study of how different groups of teachers processed visual information, described above, used slides as the stimulus materials. This focused attention on static aspects of classrooms, an artificial situation, since real classrooms are dynamic and extremely complex. To get closer to classroom realities, therefore, we designed a task that required teachers to deal with dynamic and simultaneous events (Sabers, Cushing, & Berliner, under review).

In this task a videotape of a junior high science classroom was made and then edited into three videotapes. Each tape focused on different student groups within the classroom, thus providing the effect of "seeing" the entire classroom at once. A small sample of subjects from each of the three pools we had created (experts, novices, and postulants) were shown all three videotapes *at the same time*. The subjects were asked to respond to questions about management and instruction, as well as to complete a semantic differential with respect to impressions about the classroom environment, students, and instruction. During the second viewing of the same three videotapes, shown simultaneously, subjects were asked to think aloud. After the second viewing, subjects were asked to respond to specific questions pertaining to management and instruction. Finally, a measure of subjects' memory for specific details was administered.

Interesting differences between the three groups of subjects have been found. For example, in comparison to experienced/expert teachers, novices often made inaccurate assumptions or provided contradictory statements about what they had observed, especially when they were asked about instructional or management events within the classroom. Novices experienced difficulty in making sense of their classroom observations and in providing plausible explanations about what was occurring within the classroom. For example, when novices were asked to describe the learning environment in the classroom we found:

Novice: It looked . . . I wouldn't call it terribly motivating. It was, well, not boring, but not enthusiastic.

Novice: Very positive as well as relaxed. Very positive . . . it's good to be able to focus [student] energy into a group situation, yet, at the same time, accomplishing the work that they need to do for the class and also lending to the relaxed feeling of the classroom.

Such contradictions were common. Even more discrepancy was noted when participants were asked to describe the students' attitudes toward this class. For example:

Novice: It didn't look like it was a favorite class for most of them. One boy looked kind of like, "Oh no, it's not this class again." They didn't look overwhelmingly enthusiastic to be there.

Novice: They seemed pretty excited about the class, excited to learn and a lot of times it's hard to get students excited about science, but this teacher seems to have them so that they are excited about it. They're willing to work and they want to learn.

As a group, novices seemed unable to make sense of what they saw. They experienced difficulty in monitoring all three video screens at once. Thus, they often reported contradictory observations and appeared confused about what they were observing and about the meaning of their observations.

Perhaps because postulants are even less familiar with classroom events than novices they often appeared even more overwhelmed than the novices. Many of them expressed difficulty or an inability to monitor all three video screens at once. Generally, postulants appeared able to focus on, and make sense of, only one video screen. Since this limited their observations they also made errors and contradictions when they were asked about specific events.

Experts, on the other hand, did not demonstrate confusion or difficulty in making sense of their classroom (video) observations. They felt comfortable both describing what they observed and interpreting events in terms of classroom instruction and management. For example, during the think-aloud portion of the second viewing one expert commented:

Expert: Left monitor again . . . I haven't heard a bell, but the students are already at their desks and seem to be doing purposeful activity, and this is about the time that I decided they must be an accelerated group because they came into the room and started something rather than just sitting down or socializing.

In fact, the students in the left monitor did begin working as soon as they entered the classroom and continued working throughout the entire instructional period. To us, as well as to the experts, this group of students seemed to exhibit a lot of internal motivation. Further, just as this expert noted, this was an accelerated group. It was a science classroom for students identified as GATE (Gifted and Talented Education) students.

In addition, experts were able to monitor the sounds from both the teacher and students and to use sound to assist them in understanding, interpreting, and evaluating classroom events. This use of sound allowed experts to question both their viewing of classroom management and instruction and the editing of the videos. As one expert commented:

Expert: I'm not sure if that was her comment or not about working in the lab today, but if it is [the lab], it's rather a misnomer . . . If you're listening to her comments in the background it's difficult to interpret the situation, but it's interesting that the tone of voice and the sound of the words that you hear, doesn't sound all that supportive and friendly . . . I wasn't sure if it's the editing but I'm surprised to see how quickly the students on the left had all their materials out and it seemed like they got those things without any prompting on the teacher's part . . . Listening to the comment that's on the right, the thought that goes through my mind is, I don't understand why the students can't be finding out this information on their own rather than listening to someone tell it to them.

Given their ability to monitor three video screens for both visual and auditory cues, experts seemed less confused than either novices or postulants and were better able to interpret classroom events.

We were impressed, particularly, with how experts attempted to interpret and to evaluate the events and behaviors that they described. Some illustrative excerpts from expert protocols follow.

Expert: In the monitor in the middle, it might have been a good idea to start out class with measuring the height of these plants that they're growing while roll is being taken, so that you're not wasting or having a bunch of dead time at the end of the class.

Expert: Again, viewing the middle monitor, I think there is an indication here of the type of structure of this classroom. It's pretty loose. The kids come in and go out without checking with the teacher.

Expert: I'm looking at the left monitor . . . I think that this is a part of a continuing activity from the day before, probably, because they know exactly what they're doing without any instructions from the teacher.

Expert: On the left monitor, the students' note taking indicates that they have seen sheets like this and have had presentations like this before; it's fairly efficient at this point because they're used to the format they are using.

In contrast, postulants and novices usually gave a step-by-step account of what was happening, as though they were announcing what they were viewing to someone who could not see the screen. In this respect they were reminiscent of radio announcers reporting an athletic event. The comments of postulants and novices indicated that, perhaps, they had difficulty in interpreting events and behaviors. Lacking in their "think alouds" were the inferences, conclusions, evaluations, and suggestions that appeared with frequency in the protocols of the experts.

We can generalize from these two tasks that dealt with visual information in classrooms, both as static and as dynamic information to be processed. We might expect experienced/expert teachers of mathematics to show more correct recognition of visual patterns in classrooms, more agreement about what is important, more evaluation of the events seen, more hypothesizing about what is taking place, and more accurate interpretations of events. Novice and postulant mathematics teachers would be predicted to show less of these characteristics. With regard to training mathematics teachers we see how inexperience among the novices is a source of their confusion and their lack of agreement among themselves. Such inexperience is probably the reason we also find a lack of complexity in their interpretations of visual data. Experience, probably experience that has been reflected upon, is likely to change that. Perhaps coaching or mentoring programs for new mathematics teachers can increase the speed of change from novice to experienced/expert teacher, but we are cautious about such predictions. It is more likely that significant amounts of experience as a classroom teacher are needed to acquire the characteristics we noted that our expert teachers possess.

Novices in both studies were often overwhelmed with the complexity of classroom life—its multidimensionality, simultaneity, rapidity, diversity, and so forth. Such data suggest that those who come into mathematics teaching through the alternative certification route are likely, often, to be overwhelmed, at least for an extended period of time. We would hypothesize, as well, that novice and postulant mathematics teachers would be more likely to simplify the complex environments in which they find themselves by engaging in very structured mathematics instruction, particularly emphasizing large group instruction. In such settings the chance of simultaneous events is minimized, predictability is maximized, and the pace of events is more likely to be controlled by the instructor. Expectations for creative, open-ended, process-oriented, student-led, exploratory mathematics instruction in any but the classrooms of experienced/expert teachers is likely to meet with disappointment.

The Student Information Study

For this task, each subject was presented with the following scenario. Five weeks into the school year he or she had been assigned an additional class to teach. The previous teacher had left abruptly, and her classes were being distributed among existing staff members. Experts and novices who were science teachers were given the assignment of teaching a biology class; those who were mathematics teachers were assigned an Algebra I class. Postulants were assigned at random to the science or mathematics simulation. Subjects were given a short note left by the previous teacher, a grade book with grades and attendance recorded, student information cards containing demographic information on one side and teacher comments about the student on the other, corrected tests and homework assignments, and the textbook. Participants were given 40 minutes to prepare and to write a lesson plan for the first two days of instruction. They were instructed that they should do no more or no less than they would actually do to prepare for the class. In addition, subjects were encouraged to take notes that might enable them to recall general and specific information about the class and individual students. All subjects were observed through a one-way mirror as they planned.

Subjects from each of the different subject pools were asked to recall general and specific information about the students and to make generalizations about instruction, management, and classroom organization based on the information provided. Subjects were also requested to explain their lesson plans and to select or develop a seating chart for the class. The protocols obtained from these questions provided the basic data in the study (Carter, Sabers, Cushing, Pinnegar, & Berliner, 1987).

Among the findings of interest in this study were that experts, novices, and postulants differed in their attitudes about the student information they were given. The most notable contrast was between experts and postulants. Experts were considerably less interested in remembering specific information about students than were postulants. Moreover, experts did not trust information left to them by a previous teacher. When experts did react to student information cards, it was at a comparatively more general level. Experts seemed to merge information about students into a "group picture" that they defined as more or less "typical," "normal," or "usual." Some illustrative excerpts from expert protocols follow.

Expert: I just don't think names are important as far as this point in time. I haven't met the kids; there is no reason for me to make any value judgments about them at this time. And so she [the previous teacher] had a whole little packet of confidential material that I looked at, and it had trivial little things about where the parents worked and this kid was cute or something like that, and that to me is not relevant.

Expert: Especially when I start fresh, I start from a clean slate. I usually always try to . . . I like getting a little background on the student in that there are going to be severe problems or someone may need special attention on certain things, you know, learning areas, but, in general, it's a conglomeration of the students. I like to learn from them and develop my own opinions.

Expert: It was a typical classroom, some problem kids that need to be dealt with. And you have to take that into consideration when you're developing some kind of plan for them. There are the bright kids that were highly self-motivated. There were your shy kids. It was a typical class.

Expert: I didn't read the cards. I never do unless there's a comment about a physical impairment such as hearing or sight or something I get from the nurse. I never want to place a judgment on the student before they start. I find I have a higher success rate if I don't.

Postulants' protocols indicated much more serious attention given to information about individual students. They appeared to use individual student information to slot students into two or three categories of students. Their information-processing model appeared to be one in which individual student information was useful for mentally grouping students. Examples from postulant protocols serve to illustrate.

Postulant: I sorted the bad kids from the good kids from some of the ones that were just good natured, if they liked to work, that type of thing. And I would do that if I started writing my own comments. If I had the class for a while, I would tend to still categorize them.

Postulant: I went through her student cards and also went through the test scores and tried to divide the students into three groups, one group which I thought might be disruptive, one group which I thought would not be disruptive and that wouldn't need intense watching. The third group I really didn't know because the back of the card was blank. So it was classified later. I realized not all the disruptive students were getting bad grades. I decided to sort of rank cards from what I thought would be the best student from top to the obviously poorer students going down the stack just to get some sort of an idea of ranking.

Postulant: Thirty some people I'm not going to remember right off the bat, but I'm going to try to identify who these people are. You know, your top side and your bottom side.

Postulant: I went over their test scores and looked up the ones that had low scores.

While postulants expressed this clear preference to use student information for categorizing students, their rationales for doing so were often sparse or nonexistent. As one postulant indicated:

Postulant: It seemed like the thing to do [he laughs]. I get lost . . . yeah, it just somehow seemed like the thing should be done. I'm not really sure why.

Novice teachers resembled experts in their unwillingness to focus on student information as they planned to take over a new class. However, their rationales were comparatively less rich and less specific than those of experts.

Experts, novices, and postulants in this study also differed in their routines for getting to know students and for assessing what the students had learned. Experts, more than the other two groups, saw the problems of getting to know the students and of assessing what they know about the content as two parts, and they had routines for integrating the two parts. For the expert teacher, the issues were: "What have the kids learned so far?" and "This is a new start; we need to become acquainted." All of the experts had a sense that this was a new beginning and indicated that they would begin by introducing themselves and by discussing their expectations. In response to the directive, "Tell me about your lesson plan for the first day of instruction," experts said:

Expert: The first thing that came to mind was the organization matters. I have to be organized before I feel comfortable. It can't be chaotic; I can't run a class when I don't know what's going on. I've got to be with it, what's going on, and have a routine set up the kids respond to, when they know what's expected . . . It's like training a 2-year-old; they have to know what is expected. So I put a lot of emphasis at the beginning of the year training the students.

Expert: Well, day one, I just wanted to meet the kids and find out who they were. I would lay down what my rules are, what my grading system would be, and my policies. Then I would need to question them to find out what chapters they had covered in this book. And then what I wanted to do was verbally review with them and ask questions to find out the techniques they've been taught or how they solved their greatest common factor in solving equations. From there I was going to give them all review sheets that I made because I wasn't sure where it was they had left off with.

Expert: On day one I would go in and introduce myself and maybe give an activity with the kids, a short one where they would get to know me a little better. I'd go over the class rules and assertive discipline with them. Then I'd go over their requirements for grades since most kids in high school are interested in grades.

Expert: Okay, the first day I had in mind that I'm going into a brand new class, they possibly don't know me at all. I would quickly go over what I would expect for rules and expectations, raising their hand before talking, courtesy to teacher and students . . . I'd give them some rules to follow to get us started.

In these excerpts, routines for beginning the class are apparent. The protocols suggested that, as quickly and as efficiently as possible, experts wished to "start over." They indicated they would make it clear that "A different teacher is in charge."

The novices saw a need to introduce themselves, too, but it was not with the same sense of a new beginning. The following novice protocol serves to illustrate:

Novice: I would first introduce myself and then give a brief history of myself, and I don't know it depends on the class, but I might have them introduce themselves. Then I would ask them what their prior teacher expected from them and how she ran the class and what they liked and disliked about it.

Postulants indicated they would introduce themselves to the class and would talk about the rules for the class, but they were in a much bigger hurry to get back to the book and teach what needed to be taught. As a group, they appeared to be more interested in where the students were in the book than in assessing what the students might have learned during the previous weeks. For example:

Postulant: When I was scribbling out some notes to myself, trying to figure out where they were in the book, I guess my biggest problem with the lesson plan was to figure out where they are now and where they were going.

This narrative provides an insight into how this postulant viewed the class. He was willing to continue from where the other teacher "left off." A theme of a "new beginning" was not apparent.

When attempting to assess what the students had learned so far, novices generally attended to surface features of the problem (e.g., "I noticed first off they were involved in factoring"). Common ways to handle the problem of assessment were to ask the students where they were or to review content with them. However, their reviews focused more on ways to provide the students with information than with eliciting information from them. In contrast, the experts wanted the students to provide the information about what they knew or could do. For the expert teachers, the students would be answering questions and working review exercises primarily to assess student knowledge of the subject matter. One expert indicated he would not use any specific plans for the first two days of instruction. He explained that he would use these days to interact with the students and to assess what the students remembered to give him "a fairly good idea of what had happened the first five weeks."

The Micro-Teaching Study

Another study, currently underway, is of actual teaching performance of the experts, novices, and postulants. The method is as follows. The subjects are told that they will have 30 minutes of planning time to teach a lesson on Pascal's triangle. They are given information about the background to Pascal's triangle and how it is used in determining probabilities for binomial events. A difficult lesson was chosen because we already had some information that, with relatively easy lessons and clear cut objectives, experts and novices do not show many differences in their ability to teach. The lesson on probability seemed fair to both our science and mathematics teachers, since the teaching of genetics in biology requires probability. At the end of the 30-minute planning session the subjects were asked to discuss their plans and the activities they would use when they entered the classroom. The subjects were also given information from a pretest so that they would know the approximate range of knowledge about probability among the students they would be teaching. The subjects were also provided with some information about some of the students whose behavior might be noteworthy. Materials, such as dice, pennies for tossing, overheads showing Pascal's triangle, and so forth, were available for them to take into the teaching session. Following the planning session and the discussion of it, the subjects entered a classroom and taught a 30-minute lesson to about 15 students. The lesson was videotaped. Some of the students in this micro-teaching simulation had particular roles to play. One was designated the "groomer," spending most of her time putting on make-up and combing her hair. Another was designated the "magazine reader," who took out a magazine shortly after the lesson began and proceeded through it. Another was designated the "confused" student who, toward the end of the lesson, would raise her hand and say, "I'm confused." Another student played the role of someone who deliberately tried to take the teacher off task once or twice during the lesson. Still another student came in late and then at a later stage used the pencil sharpener during the lesson.

Following the lesson the videotape was replayed for the subject and procedures for stimulated recall were followed. The subjects were asked to think aloud during the entire showing of the tape. When the stimulated recall was over photographs of the students were presented to the subject and their impressions of the students were obtained. Two weeks after the micro-teaching experience the subjects were asked to remember what they could of the experience: what they found salient, which students they remembered, and so forth. All the subjects' responses were audiotaped and then transcribed. The typed protocols are used for analysis.

The data from this study are not yet analyzed, but a few anecdotes are in order. First of all, the artificiality of the teaching situation did not seem the least bit of a problem to the novices and postulants. Almost all our experts, however, found the situation extremely stressful. In trying to understand why they were so unhappy, we have had to hypothesize about the sources of their expertise. We learned that they would never teach a lesson as complex as one on probability with Pascal's triangle without elaborate preparation. We gave them 30 minutes to prepare. Some of our experts thought they would need 3 hours or even 3 days to adequately prepare to teach such a lesson. Almost all of them claimed that they overprepared for new teaching tasks. All who teach know that sometimes we enter classrooms relatively unprepared and we must then improvise. These experts, apparently, do not engage in such improvisations as often.

The second part of their frustration was in not knowing the students. The experts had all created well-running classrooms. Even though they may not have known all their students well, the routines they had for managing their classrooms were well understood by their students. The mild behavior problems that they encountered in this micro-teaching situation were problems they never saw more than once or twice in the course of a year in their regular classrooms. Thus, well-trained students is another source of their expertise. The combination of not knowing the material well and not having the students well-trained made these experts feel that they were losing control. We think that if they had known the content and not the students they would have felt confident. Or, even if they did not know the content, if the students had known their classroom routines the experts would have felt confident. But removing *both* sources of control from them undermined their confidence in their ability to perform. These experts are very proud people. Their teaching is usually successful. They did not like the situation we had created for them, though it certainly was not our intention at all to develop an experimental task that would produce such negative affect from them. Pilot work had been done and this issue had not surfaced.

Again, though still anecdotal, we get the feeling that our postulant teachers simply did not notice certain behaviors in the students. They almost seemed to screen them out of consciousness. They paid attention to the students who were paying attention to them. Some of the novices and some of the experts clearly saw and addressed the mild behavior problems we had built into the lesson. Few of the postulants seemed to know how to deal with these issues, and therefore, became blind to them.

When the analysis of this study is completed we should be able to talk about differences between experts, novices, and postulants in planning behavior, in actual teaching performance, in thoughts about teaching during stimulated recall, in attributions made about the success or failure of their lessons, in student evaluations of these teachers, and in delayed recall about the teaching that they performed.

SOME EMERGING THEMES

The preceding accounts are a description of the research program that is ongoing. It will be another year or two until all the data are analyzed and we learn if some common themes from these studies can be identified. Nevertheless, a few ideas about how this research fits in with other research are now apparent.

First, we believe that several of the findings from the student information study are similar to findings obtained by others. Expert teachers in that study were reluctant to give the student information cards they had any more than a cursory glance. They seemed to have no need to know what the previous teacher had to say about individual students. The expert teachers stated that the class seemed "typical" or about what they would expect the class to be. They seemed to have a sense of what kinds of students they would be teaching. Their sense of what to expect must be based on their experience of instructing numerous students. This finding apparently replicates one by Calderhead (1983). In Calderhead's study of experienced, student, and novice teachers he noted that experienced teachers had amassed a large quantity of information about students and that, in a sense, "they seemed to 'know' their students before they met them" (Calderhead, 1983, p. 5). This storehouse of information that experienced teachers have accumulated about students appears to enable them to characterize what kinds of learning and behavior problems they can expect a new class to have. In contrast, novice and postulant teachers did not demonstrate the same familiarity with students and appeared uncomfortable in trying to make predictions of what could be expected and what instructional strategies might be appropriate. These characteristics also were found in the simultaneity study.

Another finding from the student information study which concurs with previous findings concerns the issue of how teachers approach the task of beginning a new class. Many expert teachers in this study explained that they would begin instruction by concentrating on establishing rules and routines. Research suggests that effective classroom managers in elementary and junior high school classes are especially skilled in establishing rules and procedures at the beginning of the year (Emmer, Evertson, & Anderson, 1980; Evertson & Emmer, 1982). Expert teachers in this study apparently knew from their own experience that familiarizing students with their routines, rules, and procedures was initially more important than introducing the students to subject matter.

In two of the studies the importance of "typicalness" or "ordinariness" came up in analysis of the expert's protocols. In the slide task it was seen in their perception of classrooms. In the student information task it was seen in their not thinking much about individual students or recalling many details about them. Experts, apparently, have images of how things ought to be. If students or classrooms appear to be the way they are supposed to be, they are likely to be less attended to—perhaps even ignored. Experience seems to change people so that they literally "see" differently, either by noting atypicality more quickly or by simply not seeing certain ordinary things. Surely that is functional. In any domain of expertise one must learn through experience, perhaps because of the severe biological limits humans have for processing information simultaneously.

Postulants in three of these studies showed an opposite characteristic: an inability to determine what was important. In the slide study they perceived widely disparate characteristics of the visual field. Wall charts, attending behavior, and the trees outside were apparently all given equal weight in terms of their salience or importance. In the simultaneity study the postulants were confused, and unable to differentiate important from unimportant events. In the student information study, the postulants noticed scribbles on tests, whether homework was dated, and other unimportant attributes, along with some of the more important attributes. Apparently, everything also had meaning for them when they examined the student information cards. They struggled to learn all they could while the experts often casually flipped through the cards. In these studies, when it came to knowing what was or was not important, novices resembled the postulants more than the experts, although they were more varied.

Experts in many situations gave responses showing a work orientation. Neither the postulants nor the novices seemed to use the term *work* as much as the experts. The postulants and novices were not guided in their thinking by notions of classroom work, teaching/learning activities for accomplishing work, control of attention to get work done, and so forth. This is a theme we hope to explore further.

In many of the situations studied so far, the experts showed the ability to relate what they saw or read to their own easily accessed rich histories of classroom experience. Episodes were often recounted about personal (often successful) experiences with a student like the one in a slide, in the class records, or in the video. Situations or problems that came up in the tasks were related to similar real-life instances in highly relevant ways. Rich images, prototypes, or schemata for students and classroom events seemed to be behind many of the responses made by the experts.

The experts in many of the tasks we studied provided evidence of more reasoned thinking than did the novices or postulants. They demonstrated this in the slide study by making more statements that cited evidence. They often reported that "because of X I assume Y." In the student information study, they showed more "if-then" thinking. Romantic images of the teacher as something like an "hypothesis maker" have often appeared in the literature on teaching. A bit of evidence for just that kind of reasoned thinking is demonstrated by the experts we study. On the other hand, novices and postulants seemed to give more responses that were unsubstantiated by evidence or to give more unjustified opinions.

Finally, the experts in the slide study and the simultaneity study showed a certain consistency in what they chose to comment about, and the comments that they made showed considerable agreement. Postulants, and to a lesser extent novices, often made statements that showed disagreement; that is, their comments were often contradictory. Furthermore, what they chose to comment about was more variable than were the choices made by the experts.

If we generalize to mathematics teaching we might expect experienced/expert mathematics teachers to have decided notions about their students: what they are like, what they can do, and what they cannot do in mathematics. The experienced/expert mathematics teacher is likely to have various and elaborate schemata about different aspects of students and their school-related behavior. Moreover, the schemata are easily brought into focus; they appear to be readily available for the experienced/expert teacher. This does not seem to be the case for novice mathematics teachers or for those who enter teaching by an alternative certification route. These well-developed schemata probably provide the cognitive backdrop for the experienced/expert teachers' frequent assumptions and hypotheses about what is seen and happening. The experienced/expert mathematics teachers are more likely to have a working management system in place, with clear roles and expectations communicated to students. Routines are likely to be used more frequently and with greater success in more areas of instruction. The experienced/expert mathematics teachers are also more likely to require more preparation time than a novice or postulant for teaching new or difficult subject matter. They are likely to be more easily thrown off than either novices or postulants when they perceive that their control of the classroom is reduced.

CONCLUSION

In this ongoing research program our strategies have been very eclectic. Such eclecticism is appropriate at the start of a research program where many difficult-to-solve problems are encountered, each of which could become an impediment to studying the phenomena of interest. Freewheeling experimentation and ignoring of problems can be tolerated, for a time, in order to explore a new research area. But eventually problems of a conceptual nature and problems in design of research studies must be addressed. Two such problems have been inadequately dealt with in our program of research. The first has to do with our attempts to understand what domains of knowledge are used by expert teachers in accomplishing their tasks. To study expertise in physics or mathematics, one studies the experienced individual's solutions to physics or mathematics problems. It is assumed that there is a one-to-one correspondence between the knowledge of, say, mathematics, that is possessed by someone and the effectiveness with which that person solves mathematics problems (Chi, Glaser, & Reese, 1981). Classroom mathematics teaching, however, seems to be much more complicated. It would seem to require that other domains of knowledge, besides content knowledge, be brought to bear if the problems of classroom mathematics teaching are to be dealt with effectively. Some have argued that a personal knowledge of self is the key requirement (e.g., Lampert, 1984), and others have identified dozens of domains of knowledge that are drawn upon by teachers (e.g., Elbaz, 1981). This is a difficult conceptual problem when attempting to determine the sources of expertise in teaching.

Our solution was simply to stipulate that there are two separate domains of knowledge that require blending in order for expertise in teaching to occur. These are: (1) subject matter knowledge and (2) knowledge of classroom organization and management, which we call pedagogical knowledge. This conceptualization is neither unique nor complex. We feel it is probably inadequate, as well, although it is a temporarily useful way of thinking. Researchers, curriculum theorists, and philosophers of education have not yet dealt adequately with the issue of what domains of knowledge are of use to the successful classroom teacher. Such knowledge would be very useful to have.

It is noteworthy, nevertheless, that even this simple conceptual framework, wherein two general knowledge domains are stipulated, has an important implication. The impliation is that teaching mathematics, for example, in contemporary U.S. public schools, is so complex in comparison to being merely a practicing mathematician who solves mathematical problems, that at least *two* complex and extensive knowledge domains must be used at all times. To be an expert teacher of a subject matter is, we think, far more complex than to be an expert problem solver in that subject matter domain.

To simplify our work, however, we did not attend too much to the content or subject matter knowledge domain. It was the pedagogical skills, management, organization, processing of classroom information, conceptions of students, and so forth that were the primary foci of our inquiries. Perhaps future researchers could find ways to study both content and pedagogy simultaneously. We clearly recognize from our studies the need to move on, eventually, to studies of expert and novice teachers where content knowledge becomes the focus of the research as well.

We have already noted that experience and expertise in many of these studies are confounded. This is both a conceptual and a methodological problem. It is a conceptual problem in the study of teaching because mere experience is not believed by most people to correlate highly with expertise in teaching. This is not as true in other fields such as radiology, physics, or air-traffic radar control. Perhaps it is because pedagogical problems are ill-structured, with no apparent "right" solutions or "winning" strategies. Perhaps it is because we have not easily agreed upon outside criteria of expertise as in chess or bridge playing. Thus, the definition of expert is always open to question and may or may not rely on knowledge and skill gained through experience. But whatever the reason, the problems of studying expertise in pedagogy are harder than in some other fields because of the widespread belief that we need to separate the concepts of expertise and experience (despite their having a common root word and their being once nearly synonymous).

Methodologically we can deal with this issue more easily than we can conceptually. We recommend that future researchers, in their experimental design, include as a group of subjects those who are experienced but not accorded the status of experts by any of the usual criteria. In comparison with the attributes of experts and novices, the role of experience can then be evaluated. Some longitudinal studies will be needed, eventually, to study how and in what ways experience changes some classroom teachers without necessarily turning them into experts. The experiences and contexts that are positively correlated with the development of expertise also need to be documented. We know very little about the acquisition of pedagogical skill by teachers during their first few years on the job. Both how they acquire pedagogical knowledge and how their content knowledge is transformed by their newly learned pedagogical knowledge into "teachable" and "learnable" classroom curriculum is yet to be explored well.

Comparisons between experts and novices in education is a research area that is still quite new, and much needs to be done. But already it appears to be a fruitful line of inquiry. We do believe expert teachers exist, share commonalities with experts in other fields, and are likely to help us understand more about how pedagogical skill is acquired. Ultimately, policies about the training of teachers, the treatment of first-year teachers, and certification may be affected by research of this type. Thus, for a few years at least, comparative research of the type described in this paper would be a worthwhile investment of resources by research funding agencies.

REFERENCES

Berliner, D. C. (1986a). In pursuit of the expert pedagogue. *Educational Researcher*, *15*(7), 5-13.

Berliner, D. C. (1986b, October). *The place of process-product research in developing the agenda for research on teacher thinking*. Invited address at the Third Conference on Teacher Thinking and Professional Action, International Study Association on Teacher Thinking, Leuven, Belgium.

Brooks, & Hawke. (1985, April). *Effective and ineffective session- opening teacher activity and task structures*. Paper presented at the annual meeting of the Amerian Educational Research Association, Chicago.

Calderhead, J. (1983, April). *Research into teachers' and student teachers' cognitions: Exploring the nature of classroom practice*. Paper presented at the annual meeting of the American Educational Research Association, Montreal.

Carter, K., Cushing, K., Sabers, D., Stein, P., & Berliner, D. C. (in press). Expert-novice differences in perceiving and processing visual classroom information. *Journal of Teacher Education*.

Carter, K., Sabers, D., Cushing, K., Pinnegar, S., & Berliner, D. (1987). Processing and using informatikon about students: A study of expert, novice, and postulant teachers. *Teaching and Teacher Education, 3*, 147-157.

Chi, M. T. H., Glaser, R., & Farr, M. (in press). *The nature of expertise*. Hillsdale, NJ: Lawrence Erlbaum Associates.

Chi, M. T. H., Glaser, R., & Rees, E. (1981). Expertise in problem solving. In R. Sternberg (Ed.) *Advances in the psychology of human intelligence, Vol. 1* (pp.7-75). Hillsdale, NJ: Lawrence Erlbaum Associates.

de Groot, A. D. (1965). *Thought and choice in chess*. The Hague: Mouton.

Elbaz, F. (1981). The teacher's "practical knowledge": Report of a case study. *Curriculum Inquiry, 11*, 43-71.

Emmer, E., Evertson, C., & Anderson, L. (1980). Effective classroom management at the beginning of the school year. *Elementary School Journal, 80*(5), 219-231.

Evertson, C. M. & Emmer, E. T. (1982). Effective management at the beginning of the year in junior high classes. *Journal of Educational Psychology, 74*(4), 485-498.

Feltovich, P. J., & Patel, V. L. (1984, April). *The pursuit of understanding in clinical reasoning*. Paper presented at the annual meeting of the American Educational Research Association, New Orleans.

Glaser, R. (in press). On the nature of expertise. In C. Schooler & W. Schaie (Eds.), *Cognitive functioning and social structure over the life course*. Norwood, NJ: Ablex.

Good, T. L., Grouws, D. A., & Ebmeier, H. (1983). *Active mathematics teaching*. New York: Longman.

Hanninen, G. (1985). *Do experts exist in gifted education?* Unpublished manuscript, University of Arizona, College of Education, Tucson.

Lampert, M. (1984). Teaching about thinking and thinking about teaching. *Journal of Curriculum Studies, 16*, 1-18.

Leinhardt, G. (1985, April). *The development of an expert explanation: An analysis of a sequence of subtraction lessons*. Paper presented at the annual meeting of the American Educational Research Association, Chicago.

Leinhardt, G., & Greeno, J. G. (1986). The cognitive skill of teaching. *Journal of Educational Psychology, 78*, 75-95.

Lesgold, A. M. (1984). Acquiring expertise. In J. R. Anderson & S. M. Kosslyn (Eds.), *Tutorials in learning and memory: Essays in honor of Gordon Bower* (pp. 31-60). San Francisco: W. H. Freeman.

Rosenshine, B. V., & Stevens, R. (1986). Teaching functions. In M. C. Wittrock (Ed.), *Handbook of research on teaching* (3rd ed., pp. 376- 391. New York: Macmillan.

Sabers, D., Cushing, K., & Berliner, D. (in press). *Differences between expert, novice, and postulant teachers in a task characterized by simultaneity, multidimensionality, and immediacy*. Tucson, AZ: University of Arizona, College of Education.

Shulman, L. S. (1986). Those who understand: Knowledge growth in teaching. *Educational Researcher, 15*(2), 4-14.

Shulman, L. S. (1987, February). *A case literature for teacher education*. Paper presented at the meeting of the Association of Teacher Educators, Houston, Texas.

Stein, P., Clarridge, P., & Berliner, D. C. (in preparation). *Teacher estimation of student knowledge: Accuracy, content, and process*. Tucson, AZ: University of Arizona, College of Education.

Strauss, S., & Stavy, R. (1982). U-shaped behavioral growth: Implications for theories of development. In W. W. Hartup (Ed.), *Review of child development research*: *Vol. 6* (pp. 547-599). Chicago: University of Chicago Press.

This research was funded in part by the Spencer Foundation, Chicago, Illinois to whom we are grateful. Important contributions to some of these studies were made by Kathy Carter, Jeremy George, and Gail Hanninen.

We acknowledge the support provided by the Center for Research in Social Behavior, University of Missouri-Columbia. We especially thank Cathy Luebbering and Diana Chappel for typing the manuscript.

Content Determinants
in Elementary School Mathematics

Andrew Porter, Robert Floden,
Donald Freeman, William Schmidt, John Schwille[1]
Institute for Research on Teaching
Michigan State University

Teachers determine what is taught in school. They create opportunities for students to learn the knowledge, skills, and dispositions that influence future productivity in school and in the social and vocational worlds beyond school. Teachers influence this effect by deciding what content to teach and by implementing strategies to engage students in that content.

This proposition has served as the central hypothesis for a line of research undertaken at the Institute for Research on Teaching (IRT). This paper summarizes what has been accomplished from those inquiries. New theoretical constructions have evolved to support analyses of school content and the methods used to determine school content. These constructions and their empirical bases have proven to be powerful mechanisms to understand practice and the ways it might be improved. The constructions also serve to elevate the importance of content in research on teaching and on educational policy.

Starting With Content

Distinguishing between the content (what is taught) and the strategy (how content is taught) of instruction ensures consideration of each (Freeman, 1978). Only if instruction centers on important content does it have potential for being worthwhile. Yet, until recently, most researchers have taken content for granted, focusing their attention on methods instead (Schwille, Porter, & Gant, 1979; Schwille et al., 1979) Hesitancy to confront issues of what should be taught is understandable. Value judgments are required that cannot have their justification in empirical fact.

Distinguishing content from strategy elevates the importance of content and raises new questions. A framework that clarifies the distinction between content and strategy has evolved from IRT research. Teachers determine (a) how much *time* is allocated to a subject, such as mathematics, over the course of a school year; (b) what topics are taught; (c) what topics are taught to *which students*; (d) *when and in what order* each topic is taught; and (e) to *what standards of achievement* a topic is taught (Schwille et al., 1982). Collectively, these five decisions determine student opportunity to learn, a major influence on student achievement (e.g., Barr, Dreeben & Wiratchai, 1983; Carroll, 1963). They specify for teachers areas of content decision making that are separate from decisions about strategy. They suggest a series of questions that teachers, policymakers, and consumers of education can use to monitor the content of schooling. They form the dependent variables in IRT research on teacher content decision making.

Understanding content also requires operational definitions of topics within a content area. Elementary school mathematics serves as the focus for IRT research on content decision making. Mathematics is a basic skill learned primarily in school. Because of the many important mathematics topics and the limited amount of school time allotted for them, decisions about what content to include in the curriculum are crucial. Nevertheless, elementary school mathematics provides a conservative test of the importance of teachers' content decisions because most people believe the content to be fairly standard (e.g., fourth graders study multiplication).

A three-dimensional taxonomy to describe the content of elementaryschool mathematics provides definitions of topics that may or may not be studied in elementary school (Kuhs et al., 1979). The three dimensions of the taxonomy describe general intent (e.g., conceptual understanding, skills, applications), the nature of material presented to students (e.g., fractions, decimals), and the operation the students must perform (e.g., estimate, multiply). The terminology and specificity of the taxonomy are based largely on an interview study of content distinctions made by elementary school teachers (Schmidt, Porter, Floden, Freeman, & Schwille, in press). Specific topics are represented by the intersections of these three dimensions (e.g., story problems involving addition of fractions, basic multiplication facts, understanding the relationship between multiplication and division). More general topics are addressed by the marginals of the taxonomy (e.g., emphasis given to conceptual understanding). Because topics can be defined at different levels of specificity, because the taxonomy has a structure that makes clear both what is taught and what is not taught, and because the distinctions made reflect ways in which teachers think and talk about their mathematics instruction, the taxonomy, when coupled with the other four attributes of content decision making, provides a language to support deliberations about content by practitioners, policymakers, and researchers (e.g., Freeman, Kuhs, Knappen, & Porter, 1982; Porter, 1983a).

The results from content analyses of instructional materials illustrate the power of this taxonomy of elementary school mathematics topics. Analyses of four commonly used fourth-grade textbooks and the five most commonly used nationally normed standardized tests of mathematics achievement (at the same grade) reveal that of the 385 topics covered by at least one of these published materials, only six topics are common to all nine. Among the textbooks, 19 topics define a core curriculum on which approximately half of the exercises in each book are focused, but the other parts of the books are idiosyncratic in their topic coverage (Freeman, Kuhs, et al., 1983). The image of a national curriculum in elementary school mathematics begins to fade, and the problems of curricular validity in educational assessment begin to emerge (Floden, Porter, Schmidt, & Freeman, 1980; Porter, Schmidt, Floden, & Freeman, 1978; Schmidt, Porter, Schwille, Floden, & Freeman, 1982; Schmidt, 1983).

The Role of the Teacher: Bounded Rationality

At least in elementary school mathematics, teachers serve as political brokers in the process of content determination (Lipsky, 1980; Schwille et al., 1982). Teachers have some discretion to follow their own convictions, but they are subject to a variety of factors that bear on their content decisions. Decisions about academic content, however, are not always primary for teachers. Teachers often plan in terms of activities rather than content outcomes (Clark & Yinger, 1979); for many elementary school teachers, academic content takes second place to other goals of schooling, such as promoting good citizenship among students (Prawat & Nickerson, 1985).

In the absence of other advice, teachers are likely to follow their own repertoires and convictions. They will teach what they have taught before, what they feel comfortable with, and what they deem appropriate for their students. But teaching does not take place in a vacuum. Advice on what to teach comes from a variety of sources and in many different forms. Students and their parents can have direct and indirect effects on what is taught. Other teachers, the school principal, the district curriculum coordinator, a university professor—all serve as potential sources of advice, as do materials and position statements from professional organizations. These interpersonal and organizational influences bear directly on teachers and operate in addition to federal, state, district, and school policies. Mathematics objectives, testing programs, mandated textbooks, promotion policies, and time guidelines all address aspects of content decision making.

The teacher stands between the content messages from these various sources and the students to be taught. The effects of advice or prescription on what to teach are mediated by the teacher's own convictions about what should be taught. To have an effect on a teacher's content decisions, then, an external influence must either change the teacher's conception about what is most desirable (i.e., persuade the teacher) or override the teacher's beliefs, forcing the teacher to comply even though the request is not viewed as appropriate.Effects of both types have been found, although persuasion is clearly the dominant form (Schwille et al., 1986; Floden et al., 1986; Porter, 1983b).

Sources of Influence

An Overview of Five Studies

Two early studies of teacher content decision making in elementary school mathematics led to increased attention on school policies (Floden, Porter, & Schwille, 1980; Schwille et al., 1982). In both of those studies school policies appeared to be among the strongest influences, after the teacher's own convictions, on what is taught. Policy effects were not uniform, however, and the range of policies considered was limited. Based on that early work and previous analyses of educational policies (particularly Spady & Mitchell, 1979), a fourfold structure was hypothesized for explaining differences in policy strength.

Policies can vary in their prescriptiveness, consistency, authority, and power. Prescriptiveness refers to the extent and specificity of a policy in telling teachers what to do. A mandated textbook is less prescriptive than a mandated textbook that teachers are instructed to follow closely, starting at the beginning and carrying through to completion. Consistency refers to links among policies, and describes how policies can contradict or reinforce each other. For example, a mandated textbook may be tied to mathematics objectives through a guide that describes pages in the book on which material is found for each objective; policies can gain authority through appeal to law, social norms, expert knowledge, or support from charismatic individuals; rewards and sanctions tied to policies give them power. Five studies have been completed, each of which addresses a different aspect of teacher content decision making in elementary school mathematics and all of which provide empirical tests of the four-attribute structure for describing the strength of content policies.

The earliest study conducted in 1978, used policy-capturing methodology to investigate the effects of six possible sources of advice on teachers' topic selection: a district mandated textbook; objectives published by the district; tests with results published by grade level and building in the local newspaper; advice from the principal; advice from upper-grade teachers; and advice from parents (Floden, Porter, Schmidt, Freeman, & Schwille, 1980). Sixty-six fourth-grade teachers were asked to imagine they had transferred to a new school and were to teach a class of fourth graders capable of fourth-grade work. They were then asked how likely they would be to add five topics that they had not been teaching and how likely they would be to drop five topics they had been teaching.

A second study conducted during the 1979-1980 school year, moved the work from the controlled setting of simulations to the real world of classrooms. Seven third- through fifth-grade teachers in six schools across three school districts were studied for a full school year to determine the mathematics content they taught, the advice they received concerning what should be taught, and the relationships between the two. Content was described through daily teacher logs (collected weekly). Advice was monitored through interviews (weekly), questionnaires, observations, and analyses of district and state policies and practices, and by attending meetings with the teachers or district-level meetings at which mathematics content might be discussed. The findings from these first two studies led to the design and completion of a series of three studies focusing on the nature and effects of state-, district-, and school-level policies.

For the study of state policies (conducted in 1981), seven states were selected to represent variation in types of policies, overall strength of policies, and school populations served: California, Florida, Indiana, Michigan, New York, Ohio, and South Carolina (Schwille et al., 1986). For each state, a complete set of documents on relevant policies and practices was assembled (e.g., objectives, testing, textbooks, allocation of time, school evaluation, teacher qualifications and promotion of specific topics). Documents were identified and additional information was collected through interviews with knowledgeable persons in each state (an average of eight persons per state).

In the second of the three studies (conducted in 1982), the study on district policies, their relationships to state policies and their perceived effects were studied in five of the seven states (Floden et al., 1986). Questionnaires were used to collect information from district mathematics coordinators, principals, and teachers using a probability-in-proportion-to-student-enrollment design for each state. Questionnaires asking about the nature of policies and their perceived effects were designed along the lines of the four-attribute structure to describe policy strength.

The third and final study once again brought the work back to the classroom. The effects of state-, district-, and school-level policies were examined for 32 fourth- and fifth-grade teachers in six Michigan school districts (Porter, 1986). Teachers provided descriptions of their daily mathematics instruction during 1982-83, using teacher logs and weekly questionnaires for each of three target students (differing in perceived ability). Districts were selected to contrast types of content relevant policies; schools were selected to contrast student body socioeconomic status; teachers were selected to contrast grouping practices when teaching mathematics. Teachers were interviewed and completed questionnaires over a three-year period to provide information on their content decision making and on their understanding of school, district, and state policies and practices concerning mathematics content. District curriculum coordinators were also interviewed over the same three-year period and documents describing district policies and practices identified in those interviews were obtained so that shifts in district policy formulation over time could be monitored. Principals were interviewed at the time teacher logs were collected to determine school-level policies and practices and to understand how principals promote state and district policies and practices.

Weak Policies, Strong Effects

The five studies provide insight into the nature of content policymaking at the state, district, and school levels and the influences of those policies on teachers' practices. The picture that emerges is one of relatively weak and fragmented policies when judged against the attributes of prescriptiveness, consistency, authority, and power, but also one of increasing policy activity over time. States, districts, and schools differ sharply in their approaches to content policy formulation. New York, South Carolina, Florida, and California have policies similar to the centralized national school systems of Europe; the policies specify what to teach and to what standards, although even these states differ in the extent to which their policies appear to challenge teacher practice. The policies of other states, such as Indiana and Ohio, operate indirectly, imposing requirements on school districts without directly telling teachers what to do. For example, Ohio has no state testing program, but the state requires school districts to have their own testing programs. Some states, such as Michigan, place great trust in local school districts and in the individual classroom teachers, avoiding prescriptions about what should be taught and to what standards of achievement (although even Michigan has a minimum objectives testing program that districts and teachers may look to for guidance) (Schwille et al., 1986).

Like states, districts also differ in the breadth and strength of their content policies. A relationship, however, exists between state and district policy practices; district policy formulation is more active in states that are also active in content policy formulation. Districts tend to extend and elaborate state policies rather than fill in areas in which states have not been active (Freeman, 1983; Cohen, 1982).

At least in elementary school mathematics, policies tend to be only mildly prescriptive; neither are they carefully constructed to be mutually reinforcing (however, they do not contradict each other either). Little evidence exists that ties teacher compliance to rewards and sanctions, nor do teachers view this to be the case. Rather, policies attempt to persuade and gain their strength through appeals to authority. Involving experts (both teachers and mathematics education experts) in the formulation of policies is the most common method for giving authority to policies. Considerable attention is also given to building policy strength through appeals to legal authority, consistency with social norms, and support from charismatic individuals (Floden et al., 1986).

Because policies rely on authority more than on power, teachers' conceptions of appropriate topics to teach are generally reflected in the policies teachers adopt. Thus, unless there is a push in a new direction, even when policies are discontinued teachers tend to continue their content practices as though the policies were still in effect.

State, district, and school content policies are relatively weak (at least from a theoretical perspective); thus, their influence on teacher content practices is surprising. Virtually every teacher studied has had his or her mathematics instruction influenced in important ways by one or more school policy. Yet the effects of content policies have not standardized teacher practice (e.g., Schmidt, et al.,in press). Perhaps because the content policies are not as prescriptive as they might be, or strong in other ways, teachers interpret policies differently. For example, in one district with a management-by-objectives system for elementary school mathematics, one teacher used the system to individualize mathematics during one period of the day but also taught mathematics during an additional period using a different textbook and whole-group instruction. Another teacher used the system as a template for deciding what to teach, when, and to what standards of achievement to each of his students, allowing students to leave the system only after they completed objectives well beyond their current grade level. Yet a third teacher only referred to the district objectives occasionally when planning instruction (Porter & Kuhs, 1982).

In another district that had recently adopted a new textbook, one teacher followed the book page by page, recognizing that the desired effect of a standardized curriculum in the district would be achieved only by following the book closely. Another teacher, not recognizing the motivation behind the single text adoption, followed her own strong convictions about what content should be taught and when, using the text only as a resource for student exercises that fit her own internal syllabus (Freeman & Schmidt, 1982).

Textbooks and Tests as Special Cases

One of the myths exposed through work on teacher content decision making is that teachers teach the content in their textbooks (Porter, 1985). Elementary school teachers view mathematics textbooks as resources to be drawn from and to be added to as seems appropriate (this belief remains unchallenged even when the textbook is mandated). Further, because textbooks do not address several of the most important content decisions, their influence is limited primarily to topic selection. Textbooks contain few instructions about how much time should be allocated to mathematics or about differences among students concerning what should be taught; they offer ambiguous advice about standards to which students should be held. Even in topic selection, most teachers cover only a fraction of their textbook's content (e.g., Freeman, 1983) and spend 10% to 20% of mathematics instruction time covering topics not in the book.

Another myth exposed as being only a half-truth is that teachers teach topics that are tested. Little evidence exists to support the supposition that national norm-referenced, standardized tests administered once a year have any important influence on teachers' content decisions. There are, however, important effects from curriculum-embedded tests (e.g., tests tied to objectives in a management-by-objectives system, chapter tests in a textbook, tests developed by teachers to help make placement decisions). Tests have effects on content decisions only when they have been explicitly tied to the curriculum and when they are readily accessible and easily used by teachers (Kuhs et al.. 1985).

Student Effects

Teachers' content decisions are also influenced by students and students' parents. Sometimes the effects are direct, coming in the form of requests to cover specific topics or requests for more homework. More often the effects are indirect, coming in the form of expectations. Student and parent effects are not random: they correlate in important ways to student characteristics such as aptitude, gender, and ethnicity.

When mathematics is taught to ability groups or to individuals, within-class content differences are dramatic. Primarily, these differences concern the topics of study rather than the total amount of time spent or the standards to which students are held. Low-ability students spend far more time learning facts and computational skills, whereas students of higher ability spend more time developing understanding of mathematical concepts and applications. High-ability students cover more topics and spend less time per topic than do low-ability students (Irwin et al., 1985).

Individualized instruction shows some evidence of gender effects. Girls encounter a larger number of topics, whereas boys study fewer topics for more time. Boys study topics that involve more conceptual understanding, more applications, and more work with pictures. Some evidence suggests an interaction between perceived ability and ethnicity. Regardless of beginning achievement scores, black girls study fewer topics than do other students, as well as fewer topics related to conceptual understanding and applications (Irwin et al., 1986a).

Whole-group instruction, however, is the primary method used to teach elementary school mathematics, minimizing differences in content among students within classrooms. Further, for a given teacher, the effects of differences among groups of students across years appear to be minimal. Even when a class is judged by the teacher to be unusually good or unusually "slow," modifications to accommodate those differences are slight. The large effects of students on teacher content decision making take place at the aggregate level. The socioeconomic status (SES) of the school student body correlates with the degree of parental influence on content, the instructional resources available to teachers, the amount of time spent on mathematics, and the topics covered (Irwin et al., 1986b). In affluent neighborhoods, parents are seen as a legitimate source of advice, generally concerned with what their children are taught. In schools that serve working class or unemployed families, parents are viewed as uninterested in particular content, even lacking the understanding required to help their children. Lower SES schools have fewer resources available for mathematics instruction. Lack of rulers and protractors affects work in measurement and geometry, and limited textbook availability affects the frequency of homework assignments. High SES schools spend less time on mathematics but cover more topics than do lower SES schools. Lower SES schools emphasize more computation and less application and concept instruction.

The correlations between the content of instruction and student characteristics are problematic. There is a tension between the amount of time students need to master content and the range of content they can cover. If understanding mathematical concepts and applications is important, however, then all students deserve an opportunity to study that content. Schools and teachers must be attentive to and must manage the dilemma to provide students time for mastery as they assure them access to useful content.

The Case of the Missing Principal

In this summary of content determinants research, policies are featured because of their surprisingly strong effects and because the number and strength of content policies is increasing at both state and district levels. Principals are featured for the opposite reasons. Despite literature emphasizing the importance of principals in school leadership and in the adoption of innovations, principals are not a major influence on teachers' decisions about what to teach in elementary school mathematics.

The literature on principal leadership and this conclusion about content decision making are not necessarily contradictory. On the rare occasions when principals have attempted to exert influences on content, teachers have accepted the attempts as legitimate, and the influence of those attempts was felt in classroom practices. But most principals remain silent on content preferences, leaving content decision making to their teachers at the classroom level and to policymakers at higher levels. Even more surprising, principals show little interest in ensuring that teachers carry out district policies. Many principals have little knowledge of district policies, devoting their efforts instead to such noncontent areas as student discipline and attendance (Floden et al., 1984).

Teacher Convictions

Differences among teachers in the content of their elementary school mathematics instruction are more substantial than can be attributed wholly to differences in policies, students, principals, or other external factors. For example, teachers at the same grade level have been found to differ in their allocation of time to mathematics by a factor of 1.5 (9000 minutes versus 6000 minutes across a full school year). Of similar magnitude, differences among teachers exist concerning the average amount of time allotted per topic. Teachers agree in their emphasis on computational skills over concepts or applications, but within that emphasis, percentage of time devoted to computational skills ranges from a low of 55% to a high of 80%. At the level of specific topics, the differences among teachers are too many to summarize. Some of these differences may even out over years for students, but students with a teacher who fails to cover geometry or who gives little attention to estimation or measurement applications are unlikely to have those omissions compensated for by other teachers in later grades.

Differences among teachers in the content of their elementary school mathematics instruction are partially a function of differences in convictions about mathematics. Teachers differ in their knowledge of mathematics, in their interest and enjoyment in teaching mathematics, in their beliefs about the importance of mathematics and the most important topics within mathematics, and in their expectations for what students can accomplish. But just as content policies have been judged to be relatively weak, elementary school teachers' convictions about mathematics are also weak. Elementary school teachers are reluctant to take responsibility for content decisions and often appear unaware that they do indeed make mathematics content decisions. During interviews, teachers often said that no one had ever asked about their mathematics content before. When asked to keep content logs, many teachers expressed keen interest in the results and some planned to monitor their own instruction in future years. Clearly, most elementary school teachers do not spend much time analyzing the appropriateness of the content of their mathematics instruction. Their positions on content remain largely unexamined, by them or by anybody else.

A few elementary school teachers do hold strong convictions about mathematics, looking primarily to their own beliefs to decide the content of their instruction. But these teachers are in a distinct minority. Curiously, they are not necessarily the teachers who possess the greatest subject matter knowledge (Freeman, 1986).

Generally, elementary school teachers are willing to change their mathematics content if (a) they view the change as being not too difficult, (b) what they are asked to do is within their range of knowledge, and (c) the request adds new content and does not require them to give up content they have been teaching (a point given more attention later). In the case of textbook adoptions, teachers' willingness to try new content takes an unusual twist. Teachers tend to follow a textbook most closely during the initial year of use. Once they have become familiar with a textbook and know what it has to offer, teachers feel greater freedom to make adjustments and to introduce some of their own preferences. The inclination to drift away from the book over time might be offset by policies that specify *how* teachers are to use their texts, but such policies rarely exist.

Some Thoughts on the Curriculum

Research on teacher content decision making in elementary school mathematics has not sought to evaluate the quality of the curriculum. Nevertheless, certain features stand out, virtually demanding comment. A ubiquitous and pronounced lack of balance exists across concepts, skills, and applications. Teachers spend a large amount of their mathematics time teaching computational skills—approximately 75%. The remaining time is distributed between teaching for conceptual understanding and applications in ways that vary from teacher to teacher. Most textbooks and minimum competency or basic skills objectives emphasize computation; however, nationally normed standardized achievement tests have balance across conceptual understanding, applications, and computational skills (Freeman, Belli, et al., 1983). The lack of balance in teacher attention to conceptual understanding, skills, and applications is problematic and should be addressed. Applications are both more important and more difficult to learn than are skills. Conceptual understanding is probably of more lasting value than either skills or applications. By formulating policies that are prescriptive, consistent, and carefully tied to sources of authority, it should be possible to create a more balanced curriculum.

A second feature of the elementary school mathematics curriculum is related to the first. Just as teachers devote a great deal of time to a relatively few computational skills, they tend to cover a large number of topics in the small amount of remaining time. Seventy to eighty percent of the topics taught during a school year receive 30 minutes or less of instructional time. Many of these topics are "touched on" or "taught for exposure," receiving only 5 or 10 minutes of attention during the year. In part, this phenomenon may be explained by a similar pattern of topic coverage in textbook exercises. The practice of covering many topics, each for a little time, also may be a function of teachers' greater willingness to take on new topics in their instruction than to give up topics they have been teaching (Floden, Porter, Schmidt, Freeman, Schwille, 1980). Whatever the reasons, the elementary school mathematics curriculum is thin and appears to be getting thinner. The practice of teaching for exposure raises questions about how much instructional time on a topic is enough. Are students learning that mathematics includes a wealth of interesting topics, or are they learning that superficial knowledge (knowing just a little about a lot of different things) is somehow valuable?

A third feature of the elementary school mathematics curriculum concerns what is missing. Students are rarely, if ever, asked to formulate a problem for themselves. Instead, they are given problems to solve. Mathematics receives little attention as a discipline worth knowing in its own right in addition to being a basic skill with utilitarian value. Even the utilitarian aspects of mathematics receive too little serious attention. For example, young women's and minorities' lack of valuing of mathematics is not sufficiently challenged by information about the mathematics prerequisite to qualify for later study and for many job possibilities.

Finally, although the elementary school mathematics curriculum is second in importance only to reading and language arts, it is treated as a distant second. Only a small amount of time is allocated to mathematics instruction. A few classrooms spend an hour or so a day on mathematics, but most classrooms average much less; some average as little as 20 minutes. Teacher choice seems to be an important determinant of the amount of classroom time spent on mathematics and low averages may reflect teachers' dislike of math (e.g., Buchmann & Schmidt, 1981).

Extending the Work

Research on determinants of elementary school mathematics content has been descriptive and largely nonjudgmental. The goal was to understand why teachers teach what they teach. In addition, however, a more complete and sharply focused picture of what mathematics is taught in elementary school has emerged. When judged against the suggestions of mathematicians and mathematics educators (e.g., NCTM, 1980; Romberg, 1983) the contrasts are substantial. Reasons for the discrepancies between advice from experts and practices of teachers are found in the preceding pages. In the absence of clear advice to the contrary, elementary school teachers largely teach content similar to what they remember having been taught when they were students, content that is not overly challenging to their students or to themselves.

One important direction for future research is to move beyond description, to consider work that starts with a judgment as to appropriate content and then seeks to design an environment that will encourage and support the teaching of that content. The policy environment should be designed using the characteristics of prescriptiveness, consistency, authority, and power. Special attention should be given to building policies with authority, rather than policies that rely on rewards and sanctions for their strength. Giving policy authority is consistent with calls to strengthen the teaching profession. The effects of authoritative policies are also more enduring.

When designing experiments to change the content of instruction, teacher knowledge of mathematics must be given attention. Past work has documented considerable variation among elementary school teachers in their mathematics content practices. Nevertheless, much of this variation has little overlap with the content recommended by mathematicians and mathematics educators. Generalizing from this descriptive work of the status quo, where teacher subject-matter knowledge is not found to be a strong influence on teacher content practices (e.g., Bryne, 1983; Leinhardt & Smith, 1984), might be misleading.

Shulman's recent formulation of the role of subject-matter knowledge in teaching (e.g., Shulman, 1986) points in some promising directions by raising the question of how the knowledge of mathematics required to teach mathematics is alike and different from the knowledge of mathematics required of a mathematician. At the very least, teachers are likely to be resistant, and appropriately so, to teaching content with which they themselves are not comfortable.

Past research on content determinants suggests that calls for change in teacher content practices must do more than specify desired new content.

Teachers require help in understanding that something that has been taught must be given up when something new is added. How this can best be done is not yet clear. Teachers are much less easily persuaded to discontinue teaching content than they are to try out something new.

A second and complementary direction for future research is to study the strategies teachers use to simplify their tasks to manageable proportions. Most research looks at teaching in slices. Good descriptive work is available on what teachers do on a particular day or for a particular unit of instruction. There is less work describing teaching practices for a full school year, and what has been done has been limited to studies of a single subject. No work describes teaching from the perspective of something an individual does all day, every day, year after year for a variety of students in a variety of subjects. This macroperspective raises questions of how teachers can best allocate limited time, energy, knowledge, and interest. Teachers' solutions to these allocation problems have enormous implications for students.

Interventions designed to improve teaching rarely take into account the long-term demands of providing good instruction. Perhaps as a result, efforts to improve teaching, even when successful in the short term, typically fail to have lasting effects (e.g., Porter, 1986). Experiments to change the content of instruction must include implementation strategies for teachers. If teacher subject-matter knowledge is a problem, an intervention requiring all teachers to have a firm and complete grasp of their subject matter is not useful. Accommodations to differentiate among teachers in their subject-matter knowledge are needed—accommodations that allow all teachers to provide good instruction on important content. Similar attention is required to accommodate limitations in teachers' time, energy, and interest.

A third direction in which research might move is to consider simultaneously both the origins and the effects of teacher content decisions. The work described here considered influences on what was taught (i.e., how much time is allotted to a topic, what topics are taught, what topics are taught to what students, and to what standards of achievement a topic is taught) but did not go on to assess effects on student achievement. The argument for this narrow focus was that student opportunity to learn is an important policy output of schooling and one of several important influences upon student learning. But what remains unclear is just how important content is as an influence on student achievement (e.g., Mehrens & Phillips, 1986). Research that looks simultaneously at the quality and quantity of instruction to see their combined and separate effects on student achievement is needed. Such research could also begin to investigate the possibility of important interactions between content to be taught and effectiveness of instructional strategies. Past research has looked at methods to the exclusion of content, and more recently, at content to the exclusion of methods (see Brophy, 1986, for a good discussion of research issues having to do with methods of instruction that mathematics educators might address). The interaction between these two essential components of instruction is yet to be determined.

There are several obvious extensions of past work on content determinants that might profitably be undertaken. How does mathematics content decision making of elementary school teachers differ from content decision making of mathematics teachers in high school (where content is departmentalized and teachers are subject-matter experts). How does content decision making in mathematics differ from content decision making in other subjects taught at the elementary school level. In contrast to mathematics, reading is taught for longer periods of time using materials that are more prescriptive, while writing is taught much less regularly and typically without benefit of textbook materials. These extensions of past work could proceed quickly, since much of the conceptual groundwork has been completed (e.g., defining content, teacher content log instrumentation, identification of characteristics that give policies weight). Such contrasts across levels of education within a subject and across subjects within a level of school should go a long way toward helping to clarify the role that teachers' subject-matter knowledge plays in their content decision making. Without further empirical work, the description of content decision making in elementary school mathematics must be taken as no more than that. Generalizations to other subjects and to other levels of school remain to be investigated.

Summary

Until recently, educational research has focused attention on the strategies of instruction. Content received little attention. By distinguishing between strategy and content, and by focusing on content, a great deal has been learned about teaching practices and about the interaction between educational policies (and other external factors) and teachers' convictions. The following commonly held beliefs have been challenged:

- There is a national curriculum in elementary school mathematics.
- From the perspective of content covered, materials are interchangeable.
- What is taught in one classroom closely resembles what is taught in another classroom at the same grade level.
- Textbooks determine the content of instruction.
- Teachers are resistant to top-down calls for change in matters of content.
- Policies have their effect through the manipulation of rewards and sanctions.
- Teacher autonomy is better than central control.
- Individualized instruction is better than group instruction.

- Instruction is better when teachers make substantial deviations from commercially prepared materials.

Partly as a result of research on content determinants, publishers of instructional materials are now much more aware of and concerned about curricular validity. Similarly, schools are more concerned about issues of curriculum alignment. Those responsible for monitoring education are more aware of the need to monitor the content of instruction as well as other aspects of educational inputs, processes, and outputs (e.g., the framework for describing elementary school mathematics has served as input to the National Research Council's Committee on Indicators of Precollege Science and Mathematics Education and is under consideration by the Center for Educational Assessment of the Council of Chief State School Officers). Teacher education programs are beginning to address the teacher's role in content decision making, an aspect of the teacher education curriculum that was largely missing. Educational research, especially research on teaching, now recognizes the importance of differences among teachers in their emphases on academic content. Increasingly, research studies focus on content decision making and on the ways teachers make use of instructional materials.

Work on content policies and their effect is more recent and less visible. Nevertheless, the work points to a middle ground between two developments that seem on a collision course. On the one hand, centralized control of the curriculum is increasing. States and districts are developing policies that specify what is to be taught, to whom, and to what standards of achievement. On the other hand, there is increasing concern for the status of the teaching profession. Recommendations are being made for greater teacher autonomy and greater teacher participation in school policy formulation. But central control versus teacher autonomy may be a false dichotomy. Content policies will be persuasive to teachers if teachers are meaningfully involved in establishing those policies. Under those conditions, compliance and professional autonomy become two sides of the same coin.

REFERENCES

Barr, R., Dreeben, R., & Wiratchai, N. (1983). *How schools work*. Chicago: University of Chicago Press.

Brophy, J. (1986). Teaching and learning mathematics: Where research should be going. *Journal for Research in Mathematics Education, 17*, 323-346.

Bryne, C. J. (1983). *Teacher knowledge and teacher effectiveness: A literature review, theoretical analysis and discussion of research strategy*. Paper presented at the 14th Annual Convocation of the Northeastern Educational Research Association, Ellenville, NY.

Buchmann, M., & Schmidt, W. H. (1981). Six teachers' beliefs and attitudes and their curricular time allocations. *Elementary School Journal, 84*, 162-171.

Carroll, J. (1963). A model for school learning. *Teachers College Record, 64*, 723-733.

Clark, C. M., & Yinger, R. J. (1979). *Three studies of teacher planning* (Research Series No. 55). East Lansing: Michigan State University, Institute for Research on Teaching.

Cohen, D. K. (1982). Policy and organization: The impact of state and federal educational policy on school governance. *Harvard Educational Review, 52*, 474-499.

Floden, R. E., Alford, L., Freeman, D. J., Irwin, S., Porter, A. C., Schmidt, W. H., & Schwille, J. R. (1984, April). *Elementary school principals' role in district and school curriculum change*. Paper presented at the annual meeting of the American Educational Research Association, New Orleans.

Floden, R. E., Porter, A. C., Alford, L. M., Freeman, D. T., Irwin, S., Schmidt, W. H., & Schwille, J. R. (1987). *Instructional leadership at the district level: A framework for research and some initial results* (Research Series No. 182). East Lansing: Michigan State University, Institute for Research on Teaching.

Floden, R. E., Porter, A. C., Schmidt, W. H., Freeman, D. J., & Schwille, J. R. (1980). Responses to curriculum pressures: A policy-capturing study of teacher decisions about content. *Journal of Educational Psychology*, *73*, 129-141.

Floden, R. E., Porter, A. C., Schmidt, W. H., & Freeman, D. J. (1980). Don't they all measure the same thing? Consequences of selecting standardized tests. In E. L. Baker & E. S. Quellmalz (Eds.), *Educational testing and evaluation: Design, analysis, and policy* (pp. 109-120). Beverly Hills: Sage.

Freeman, D. J. (1978). *Conceptual issues in the content/strategy distinction* (Research Series No. 21). East Lansing: Michigan State University, Institute for Research on Teaching.

Freeman, D. J. (1983, April). *Relations between state and district policies*. Paper presented at the annual meeting of the American Educational Research Association, Montreal.

Freeman, D. (1986, April). *The strength of teachers' convictions as a determinant of student opportunity to learn*. Paper presented at the annual meeting of the American Educational Research Association, San Francisco.

Freeman, D. J., Belli, G. M., Porter, A. C., Floden, R. E., Schmidt, W. H., & Schwille, J. R. (1983). The influence of different styles of textbook use on instructional validity of standardized tests. *Journal of Educational Measurement*, *20*, 259-270.

Freeman, D., Kuhs, T., Knappen, L., & Porter, A. (1982). A closer look at standardized tests. *Arithmetic Teacher*, *29*(7), 50-54.

Freeman, D. J., Kuhs, T. M., Porter, A. C., Floden, R. E., Schmidt, W. H., & Schwille, J. R. (1983). Do textbooks and tests define a national curriculum in elementary school mathematics? *Elementary School Journal*, *83*, 501-513

Freeman, D., & Schmidt, W. (1982). *Textbooks: Their messages and their effects*. Paper presented at the annual meeting of the American Educational Research Association, New York.

Irwin, S., Alford, L., Berge, Z., Floden, R., Freeman, D., Porter, A., Schmidt, W., Schwille, J., & Vredevoogd. J. (1985, April). *Grouping practices and opportunity to learn: A study of within-classroom variations in content taught*. Paper presented at the annual meeting of the American Educational Research Association, Chicago.

Irwin, S., Alford, L., Berge, Z., Floden, R., Freeman, D., Porter, A., Schmidt, W., Schwille, J., & Vredevoogd, J. (1986a, April). *Gender and elementary mathematics: Where do differences begin?* Paper presented at the annual meeting of the American Educational Research Association, San Francisco.

Irwin, S., Alford, L., Berge, Z., Floden, R., Freeman, D., Porter, A., Schmidt, W., Schwille, J., & Vredevoogd, J. (1986b, April). *The effects of school socioeconomic status on student opportunities to learn mathematics*. Paper presented at the annual meeting of the American Educational Research Association, San Francisco.

Kuhs, T., Porter, A., Floden, R., Freeman, D., Schmidt, W., & Schwille, J. (1985). Differences among teachers in their use of curriculum- embedded tests. *Elementary School Journal*, *86*, 141-153.

Kuhs, T., Schmidt, W., Porter, A., Floden, R., Freeman, D., & Schwille, J. (1979). *A taxonomy for classifying elementary school mathematics content* (Research Series No. 4). East Lansing, Michigan State University, Institute for Research on Teaching.

Leinhardt. G., & Smith, D. (1984). *Expertise in mathematics instruction: Subject matter knowledge*. Paper presented at the annual meeting of the American Educational Research Association, New Orleans.

Lipsky, M. (1980). *Street level bureaucracy: Dilemmas of the individual in public services*. New York: Russell Sage Foundation.

Mehrens, W. A., & Phillips, S. E. (1986). Detecting impacts of curricular differences in achievement test data. *Journal of Educational Measurement, 23*, 185-196.

National Council of Teachers of Mathematics (1980). *An agenda for action: Recommendations for school mathematics of the 1980s*. Reston, VA: National Council of Teachers of Mathematics.

Porter, A. C. (1983a). The role of testing in effective schools. *American Education, 19*(1), 25-28.

Porter. A. C. (1983b, April). *Policies perceived to have impact*. Paper presented at the annual meeting of the American Educational Research Association, Montreal, Canada.

Porter, A. C. (1985). *Can a book be the curriculum*? Invited address, Textbook Conference, Council of Chief State School Officers and National Association of State Boards of Education, Washington, DC.

Porter, A. C. (1986, April). *The design of a longitudinal study to fit a cumulative program of research on teachers' content decision making*. Paper presented at the annual meeting of the American Educational Research Association, San Francisco.

Porter. A. C. (1986). From research on teaching to staff development: A difficult step. *Elementary School Journal, 86*(2), 141- 154.

Porter, A. C., & Kuhs, T. (1982, April). *A district management-by-objectives system: Its messages and effects*. Paper presented at the annual meeting of the American Educational Research Association, New York.

Porter, A. C., Schmidt, W. H., Floden, R. E., & Freeman, D. J. (1978). Practical significance in program evaluation. *American Educational Research Journal, 15*, 529-539.

Prawat, R. S., & Nickerson, J. R. (1985). The relationship between teacher thought and action and student affective outcomes. *Elementary School Journal, 85*, 529-540.

Romberg, T. A. (1983). A common curriculum for mathematics. In G. D. Fenstermacher & J. I. Goodlad (Eds.). *Individual differences and the common curriculum* (82nd yearbook of the National Society for the Study of Education, pp. 121-159). Chicago: University of Chicago Press.

Schmidt. W. H. (1983). Content biases in achievement tests. *Journal of Educational Measurement, 20*, 165-178.

Schmidt, W. H., Porter, A. C., Floden, R. E., Freeman, D. J., & Schwille, J. R. (in press). Four patterns of teacher content decision making. *Journal of Curriculum Studies*.

Schmidt, W. H., Porter, A. C., Schwille, J. R., Floden, R. E., & Freeman, D. J. (1982). Validity as a variable: Can the same certification test be valid for all students? In G.F. Madaus (Ed.), *The courts, validity and minimum competency testing* (pp. 133-151). Boston: Kluwer-Nijhoff.

Schwille, J., Porter, A., Alford, L., Floden, R., Freeman, D., Irwin, S., & Schmidt, W. (in press). State policy and the control of curriculum decisions: Zones of tolerence for teachers in elementary school mathematics. *Educational Policy*. Schwille, J., Porter, A., Belli, G., Floden, R., Freeman, D., Knappen, L., Kuhs, T., & Schmidt, W. (1982). Teachers as policy brokers in the content of elementary school mathematics. In L. Shulman & G. Sykes (Eds.), *Handbook of teaching and policy* (pp. 370-391). New York: Longman.

Schwille, J., Porter, A., & Gant, M. (1979). Content decision making and the politics of education. *Educational Administration Quarterly, 16*, 21-40.

Schwille, J., Porter, A., Gant, M., Belli, G., Floden, R., Freeman, D., Knappen, L., Kuhs, T., & Schmidt, W. (1979). *Factors influencing teachers' decisions about what to teach: Sociological perspectives* (Research Series No. 62). East Lansing: Michigan State University, Institute for Research on Teaching.

Shulman, L. S. (1986). Those who understand: Knowledge growth in teaching. *Educational Researcher, 15*(2), 4-14.

Spady, W. G., & Mitchell, D. E. (1979). Authority and the management of classroom activities. In D. L. Duke (Ed.), *Classroom management* (78th yearbook of the National Society for the Study of Education, pp. 75-115). Chicago: University of Chicago Press.

NOTES

[1]The authors were the permanent members of the Content Determinants Project in the Institute for Research on Teaching. The contributions of those who were with the group for shorter periods of time are gratefully acknowledged: Linda Alford, Gabriella Belli, Zane Berge, Michael Gant, Susan Irwin, Frank Jenkins, Luci Knappen, Therese Kuhs, and Janet Vredevoogd.

This work is sponsored in part by the Institute for Research on Teaching, College of Education, Michigan State University. The Institute for Research on Teaching is funded from a variety of federal, state, and private sources including the United States Department of Education and Michigan State University. The opinions expressed in this publication do not necessarily reflect the position, policy, or endorsement of the funding agencies.

Research and the Improvement of Mathematics Instruction: The Need for Observational Resources[1]

Thomas L. Good and Bruce J. Biddle
University of Missouri—Columbia

This paper argues that observational research in classrooms has not been utilized sufficiently to improve mathematics education. We contend that the expansion of observational research can yield better theories for understanding the learning of mathematics and other subjects and can produce more adequate models for improving teaching.

The paper begins with the assertion that the present lack of useful knowledge about teaching is due in large part to inadequate funding for educational research. Whereas earlier generations may be forgiven their unreasonable recommendations for the improvement of education, it is more difficult to forgive contemporary advocates who urge reforms without first obtaining empirical information about teaching.

We next discuss reform efforts in U.S. public education. It is possible to identify several periods when sweeping recommendations for educational reform were made in this country. Unfortunately, these often involved simplistic ideas about schools or curriculum problems and little, if any, documentation of the classroom problems that reforms were intended to address. Too often reformers claimed that all teaching was similar and that all practice needed to be reformed in a simple way. Such claims are wrong and are demeaning and unfair to excellent teachers. Many of these reforms seemed to reflect an unwillingness to view teaching as a complex, challenging, multifaceted process that is still inadequately understood. Rather, they assumed that both the problems and solutions for improving U.S. education were obvious. In addition, many of these reforms were eventually judged to have failed; but since little, if any, observational data had been collected to examine their effects in classrooms, it was difficult to say why the reforms failed.

To illustrate the problems associated with past reforms, we discuss briefly two reform efforts: discovery learning and the curricular reforms of the School Mathematics Study Group (SMSG). We also note that the cycle of reform continues today and that curriculum changes are still advocated with little attention to the effects of these reforms on classroom practice.

Observational studies of classrooms find consistently that educational problems vary among schools and classrooms, thus observational research conducted *before* curricular reform is undertaken can lead to improved understanding and the testing of alternative solutions to problems. As well, observational research is needed *during* periods of curricular reform to establish the effects of those reform efforts, if any. To illustrate the values of observational studies, we also describe two programs of classroom research: one that focuses on teacher expectations, the other an effort that produced the Missouri Mathematics Program.

Our paper closes with brief disclaimers about the limitations and opportunities of observational research and a summary of our argument.

Inadequate Funding

Comparison with funding for research for busienss and defense purposes makes it painfully clear that too little of the educational budget is spent for research or for the careful development of ideas. Collective investments in educational research (universities, public schools, state departments of education, foundations, etc.) are very limited, particularly in comparison to other areas of research. Futrell (1986) notes that the federal government includes 61 billion dollars for research and development in its fiscal 1987 budget. Of these funds, .2% are allocated for educational research! In comparison, 61.2% will go for military research, 9.3% for health studies, 8.1% for energy research, and 6.6% for NASA. The Office of Research in the U.S. Department of Education, which is responsible for federal research, presently has a budget of roughly 20 million dollars. Of these funds, only $500,000 are now said to be available to support new lines of inquiry. To put these figures in perspective, the present "fly-away" cost of a single B-1 bomber is 212.5 million dollars. If this figure seems an unfair comparison, we might note that the current cost of a single ground-launch cruise missile is over five million dollars and the cost of a single F-16 fighter is about 13 million dollars (independent of development costs). Industry, unlike public education, is willing to invest its resources in research in order to understand problems and to improve or find new products. For example, the following comments, written with obvious pride were included in the 1986 Annual report of Eli Lilly and Company, "The increasing investment in research should ensure the future strength of the corporation. In 1986, our research and development expenses increased 15 percent and totalled $427 million. We spent more than 11 cents out of every sales dollar on research. A number of new research programs were begun and others were expanded significantly" (page 2). Clearly, pious expressions about the importance of education and expressed zeal for educational reform are not matched by a willingness to pay for related educational research and development. If we are serious about improving schools, it is imperative that support for basic educational research be increased substantially.

Research that examines learning in actual classroom settings is especially lacking. Whereas earlier generations lacked the tools for serious, observational research in classrooms, those tools have been available for about a decade (Evertson and Green, 1986). As a result of their use, a good deal of knowledge has already been developed concerning the teaching of mathematics and of other academic subjects. Although this knowledge has had obvious effects on educational thought and textbooks, it has not always informed the opinions of advocates who sometimes call for educational reforms that contradict available evidence. Worse, contemporary reformers often seem oblivious of the possibility that research in classrooms can be used to check the veracity of their assumptions and the effects of their programs, and proceed, in ignorance, to urge—with no empirical foundation—programs that will cost many millions of dollars and will affect the lives of thousands of teachers and pupils. Although we by no means discourage funding for other, needed educational research efforts, it seems utterly tragic that so little funding is presently available for observational studies of the classroom processes that are the heart and focus of our educational reform efforts.

Reform and Simplistic Ideas

As Good (1986) notes, most of the reform movements in this country have concerned single variables or clusters of variables focused on only one dimension of a problem. The assumption appears to be that there is a common problem; hence, there ought to be a simple answer. Unfortunately, this logic defies common experience as well as research results. The problems of American schooling vary from school to school (Good & Brophy, 1986a; Good & Weinstein, 1986a), and even teachers at the same grade level in the same school may have different problems (Carew & Lightfoot, 1979). If some classrooms have *too much* structure and other classrooms have *too little* structure, then a simple call for a change such as more time-on-task will have uneven effects.

The logic of our argument is straightforward: if problems vary, so must solutions. However, the history of U.S. educational reform is marked by the search for simplistic answers that have little or no research base. At various times educators in this century have advocated as answers large-group instruction, small-group teaching, and individualized teaching! Or, to take another example, both direct instruction and discovery learning have been posited, at different times, as panaceas for improving education.

However, it seems clear that the simple characteristics of instruction have never predicted instructional effectiveness, although this is where many of our reform efforts have been centered. The issue is not individualized instruction or small-group instruction, but rather the *quality* of thought and effort that occur within these structures. Different classroom strategies can work; the key is quality of instruction, and to find that key we must observe classroom teaching directly.

As an example, in the late 1950s and early 1960s many educators became concerned with students' phonics deficiencies, stimulated by books such as *Why Johnny Can't Read* (Flesch, 1955). Associated concerns led to the massive investment of time and other resources to remedy this problem (training teachers in phonetics, exposing students to drill sheets that provided extended practice in decoding, etc.; see Artley, 1981; Chall, 1967). Unfortunately, today some reading critics tell us that when they visit class-rooms they see too little focus on silent reading, on inferential thinking, and on reading for pleasure and enjoyment, dimensions that many of us would call central to reading (Commission on Reading, 1985). It seems that students can now decode words, but they cannot put the words together and make inferences from them. Too often in reform movements those things that are working and those things that are important are swept out along with those that are dysfunctional. Simple-minded change models pro-posed by advocates almost guarantee new problems by overreacting to existing problems. Single variables do not define problems, and intervention programs that deal with a single factor are not likely to be very successful, especially when change efforts are not based on knowledge from research. Indeed, lack of observational findings makes it easy for educators to mis-perceive problems and to believe that complex problems can be defined simply.

Another problem with simplistic reform efforts is that we tend not to learn much from past lessons; we continue to repeat the same mistakes. For example, the National Commission on Excellence in Education's *A Nation At Risk* focuses on superficial quantitative solutions that ignore educational quality. Among others, the Commission recommends a longer school day. One wonders why. If educators are doing such a poor job with the time that they have, why should giving them more time help to improve pupil perfor-mance? Why do students who are reluctant learners at 2:00 suddenly become intrinsically motivated learners at 4:00 with the advent of the extended school day? If poor student performance in mathematics is due to a drill-oriented curriculum that presents mathematics in a fragmented way in which students often memorize algorithms without understanding them, why is making more time available for mathematics instruction helpful? In such cases, more time for drill and memory work is only going to deepen a comprehension problem, not reduce it. Similarly, one wonders why asking students to take more mathematics is a solution if there are not qualified teachers to teach the extra mathematics courses. More time is a misplaced strategy unless the quality of that time is improved.

Discovery Learning and SMSG

As the preceding discussion suggests, the tendencies to define problems without careful documentation and to generate solutions without the aid of observational data appear in all curriculum areas. However, some discussion here may be useful to suggest that mathematics educators and mathemati-cians are not immune to the tendency to view schooling in too simplistic terms.

In their discussion of various *commitments*, Dunkin and Biddle (1974) note that mathematicians and physical scientists have often advocated discovery learning because of its presumed effectiveness. These authors note that, considering that advocates of discovery have ranged from psychologists to teachers to curriculum theorists, it is understandable that the message of discovery learning is vague and inconsistent. The general theme, however, is that teachers should structure learning so that students discover generalizations and rules for themselves. However, as Dunkin and Biddle note, the discovery orientation tends to suggest what teachers are not to do but fails to specify what they are supposed to do. This is not to suggest that curriculum theorists who argued for a discovery approach did not provide some direction for informed classroom practice, but it is to stress that the advice was general and undifferentiated and hence apt to be misapplied.

The discovery orientation is similar to other strongly advocated approaches that are advocated on a basis other than demonstrated effectiveness in classrooms. Such approaches arouse passion but lead to little understanding of classroom practice (for a recent example, see Sternberg, 1985, who suggests the fact that, despite the plethora of thinking skills courses, we know little about what happens when these programs are used in classrooms). Other large-scale curriculum projects such as that of the School Mathematics Study Group (SMSG) provide examples of how good ideas are supposedly translated into school experiences but with few enduring effects. Despite the vast resources and energy that went into the SMSG effort, we know little about how its material impacted classroom experiences. This is not to suggest that these were not positive outcomes from SMSG (for example we suspect that the mathematics presented in textbooks were improved by this effort); however, it is to suggest that teaching practice in many classrooms were affected in various ways, many of which were inconsistent with what reformers advocated.

Cycles of Reform

The negligible effects of past curriculum improvement efforts should serve as a reminder that change is difficult. The cycle of general concern and the implementation of sweeping, undifferentiated solutions recur regularly in U.S. education. Unfortunately, many of these cycles have produced but weak effects. Such curriculum reforms often fail because they are not adequately supported, because they overemphasize certain aspects of the curriculum to the neglect of others, and because they do not reflect a differentiated and realistic image of teaching.

Powell, Farrar, and Cohen (1985) remind us that some of the recent proposals for curricular reform occurred in earlier times as well. For example, in the 1880s and 1890s entrepreneurial public educators tried to attract out-of-school youth into high schools and to encourage students who were in private schools to attend public schools by offering various "modern courses." Reform also occurred in the 1920s and 30s to accommodate an expanding high school population who had presumably less academic talent.

This led to the creation of additional courses that were added for less capable and less interested students. Furthermore, there was an attempt to develop an activity curriculum that would make the study of school subjects more involving and more fulfilling. The need for these additions seemed obvious to all concerned; however, reformers seemed also to believe that students placed in such courses would want to learn and that their teachers would want to teach. Powell et al. wonder why the reformers held these beliefs; "did no one think that the result would be demeaning for teachers and defeating for students?" (p. 260).

Calls for curriculum reform escalated when the Soviets launched Sputnik in 1957. The demand for better instruction in science, mathematics, and languages was quickly tied to the issue of national defense, funds were soon allocated for recruiting talented students in these areas, and a decided effort was made to improve teachers' knowledge by various programs of subsidized study. Powell et al. note that those latter reforms did lead to more thoughtful, up-to-date treatment of these academic subjects, and particularly to improved textbooks. However, when one considers the actual use made of these materials, the picture of reform is less favorable. Most studies of the 50s reform suggest that the new curricula had few effects on U.S. teachers' behaviors and knowledge. As Powell et al. note, teachers were not involved in the development of these new materials, nor did reformers devise implementation strategies that would help teachers to improve their understanding while using these materials. Perhaps most importantly, materials were not developed with the classroom in mind.

The same misguided approach is now being taken in response to *A Nation At Risk*, as educators develop new math courses but are not concerned with who will teach those courses and what experiences students should have in them. We thus still tend to reform the curriculum on the basis of shibboleths and general images without careful analytical reasoning and evidence concerning the specific teaching practices that will accomplish our goals.

The Utility of Research on Teaching

Schools are social institutions and tend to evolve over time. Thus, goals and procedures within the school will gradually evolve in response to changes in community needs, the nature of pupils served, beliefs of educators, curricular reforms, and funds available. These evolutionary changes are altogether proper; indeed, since the time of John Dewey, educators have been taught that part of the strength of U.S. education lies in the willingness of its educators to consider innovations, to extend a hearing to new ideas, to view the school as a laboratory in which new procedures are constantly explored. Moreover, there are obvious advantages to considering the school in this manner. (Not only does the innovative model open the school to new ideas, but teachers within it are constantly stimulated by the excitement of innovation.)

Unfortunately, there are also disadvantages to the innovative model for education. One of these is associated with resources. It is all very well to encourage exploration when one has extra funds, but when funds are tight one has to think very carefully before trying out an expensive innovation. In addition, as in other social institutions, the tie between innovation and effect is not always easy to detect. For example, let us consider the school that is considering a new curriculum that is touted for its ability to increase mathematics achievement. Suppose that school adopts that curriculum on a trial basis but finds that achievement gains were minimal. What went wrong? Do we conclude that the curriculum is flawed, that it was not properly implemented, that its effects in the classroom were not those anticipated, that the problems it proposed to address were not present in our school, or that its impact on the pupils of our school were not as anticipated? Lacking appropriate research, each of these is a potentially viable explanation, and yet these several explanations lead to different recommendations for future action within the school.

These disadvantages suggest a role for *research on teaching* within the general model for educational innovation. If, as we argue, observational research has the capacity for generating empirically-based insights concerning the causes, conduct, and consequences of teaching, then those insights can be used by educators to inform the decisions they make when planning or evaluating innovations in schools. Among other things, those insights can help educators to resist the enthusiasms of vendors who are trying to sell an educational product. They can also lead educators to understand why certain teaching strategies are effective with some groups of pupils and ineffective with others. And they can provide information useful for anticipating, measuring, or interpreting the outcomes of innovations. In short, we argue that schools and educators make more sensible decisions, that resources are saved and U.S. education is possibly improved, when the "normal" processes of educational innovation are supplemented by the insights arising from research on teaching.

Examples of Research on Teaching

We turn now to examples that illustrate the ways in which observational research within classrooms may have an impact, by generating information and insights about teaching that have, in turn, the potential for improving classroom practice.

Teacher Expectations

Two decades ago an experiment conducted by Rosenthal and Jacobson (1968) proved to be one of the most exciting and controversial reports to appear in the history of educational research. These investigators presented data suggesting that teachers' experimentally induced expectations for student performance were related to actual levels of student performance.

Unfortunately, secondary sources describing this study made exaggerated claims that went far beyond those made by Rosenthal and Jacobson, and the study captured the imagination of the general public. Some attempts to replicate Rosenthal and Jacobson's data (e.g., Claiborn, 1969) were unsuccessful. However, subsequent research established that in many classrooms teachers' expectations can and do affect classroom behavior (see Brophy, 1983; Brophy & Good, 1974; Cooper & Good, 1983; Dusek, 1986; Marshall & Weinstein, 1984), although it appeared that the expectation problem varied from classroom to classroom, and there was then no clear general way to use this information to improve education.

One of the weaknesses of the original Rosenthal and Jacobson study was the fact that interactions between teachers and students *were not observed*; in short, no information was provided concerning the different ways teachers treated pupils for whom they had high and low expectations. Fortunately, in the last 15 years many observational studies have examined the relationship between teachers' beliefs about individual students' achievement and teachers' interactions with students.

Brophy and Good (1970) proposed a model for thinking about this relationship:

1. The teacher expects specific behavior and achievement from particular students.

2. Because of these varied expectations, the teacher behaves differently toward different students.

3. This treatment communicates to the students what behavior and achievement the teacher expects from them and affects their self-concepts, achievement motivation, and levels of aspiration. If this treatment is consistent over time, and if the students do not resist or change it in some way, it will shape their achievement and behavior.

4. Students for whom teachers' expectations are high will be led to achieve at high levels, whereas the achievement of students for whom teacher expectations are low will decline.

5. With time, students' achievement and behavior will conform more and more closely to the behavior originally expected of them.

In subsequent work (for example, Brophy & Good, 1974; Cooper & Good, 1983; Good, 1981), it has become clear that individual teachers react differently to differences among students. It has been estimated on the basis of many studies that perhaps one third of teachers behave in ways that sustain the poor performance of low achievers. Ways in which teachers express low expectations toward students whom the teachers perceive to be low achievers have been presented at length elsewhere (Good & Brophy, 1987), however specific findings suggest two different types of expectation problems. Teachers who criticize low achievers for incorrect responses seem to be intolerant of these pupils. And teachers who reward marginal (or even wrong) answers are excessively sympathetic and unnecessarily protective of low achievers. Both types of teacher behavior illustrate to students that effort and classroom performance are not related (Good & Brophy, 1986b), and both discourage student thinking.

Passivity models. Differences in the way teachers treat low achievers over time may reduce the efforts of low-achieving students and contribute to a passive learning style. To illustrate, low-achieving pupils who are called on frequently one year (because the teacher believes they need to participate if they are to learn), but who fihd that they are called on infrequently the following year (because the teacher does not want to embarrass them), may find it confusing to adjust to different role definitions. Ironically, those pupils who have the least capacity to adapt may be asked to make the most adjustments as they move from classroom to classroom. Thus, greater variation may appear in between-teacher treatment for low achievers than for high achievers because teachers agree less about how to respond to pupils who do not learn readily.

In addition, passivity may also be encouraged within the specific classroom. When teachers provide fewer chances for low achievers to participate in public discussion, wait less time for them to respond when they are called on (even though these students may need more time to think and to form an answer), criticize low achievers more per incorrect answer, and praise them less per correct answer, passivity can also result. It seems that a good strategy for students who face such conditions would be not to volunteer or not to respond when called on. Students are discouraged from taking academic risks under such an instructional system (Good, 1981). To the extent that students are motivated to reduce risks and ambiguity—and many argue that students are strongly motivated to do so (Doyle, 1983)—students would likely become more passive in order to reduce the risks of critical teacher feedback.

Good, Slavings, Harel, and Emerson (1987) have studied students' self-initiated questions in an attempt to determine whether or not "high- and low-potential" students learn different questioning skills. Findings indicated that students in grades K-12 asked similar numbers of questions. Questions about explanation were relatively infrequent, and procedural questions were relatively high, at all grade levels. However, students perceived by teachers to be low-achieving asked more questions than did high-achieving students at the kindergarten level. In contrast, as age increased, low-achieving students asked fewer questions. Good et al. note that their data lead to an instructive but distressing hypothesis. The ways in which low-achieving students present themselves through asking frequent but seemingly inappropriate questions may inadvertently lead to teacher and peer feedback that in subtle ways undermines the initiative of these students over time.

Eventually these students learn that it is better to avoid academic responding than to risk indicating that they do not understand.

Complexities of application. As these results suggest, research on teacher expectations has suggested several models for how teachers might communicate low expectations to students in classroom situations. However, it is clear that this information is useful largely as a framework for thinking about classrooms rather than as a general definition of how teachers should behave in a classroom. To illustrate, in a recent case study that focused on expectations, Good and Weinstein (1986b) found few examples of teacher behaviors that would encourage low expectations on the part of individual pupils. Instead, within the classrooms studied the major problem was the projection of low expectations to entire classes! Thus, it would appear that there are many ways in which teachers can express low expectations.

Good and Weinstein argue that teachers must be decision makers and apply knowledge from research on expectations uniquely in their own classrooms. This is because of the complex and difficult ways that teachers may respond to expectations. Expectations are expressed variously—through choice of curriculum, rationale given to students for curriculum topics, and performance feedback, among other behaviors. Thus, it is not possible to identify a single combination of behaviors that can lead to the communication of appropriate expectations. It is also difficult to provide advice to teachers about the utility of a particular teacher behavior because students' interpretations of that behavior depend not only on the behavior itself but also on the context in which it occurs. To illustrate, Brophy (1981) notes that praise may either erode or enhance motivation, depending on context, and Mittman (1985) reports that under some conditions students may interpret teacher criticism favorably (i.e., teacher cares or expects good work). Again, questions that are challenging to a sixth grader may be threatening to a third grader.

Despite these complexities, knowledge from expectation research can certainly be used to prevent silly but tragic policy decisions. Ample evidence indicates that some teachers call on low-achieving students less often than students they believe to be more capable. However, it makes little sense, to encourage *all* teachers to call on low-achieving students more frequently simply because *some* teachers call on low-achieving students too infrequently. Similarly, it does little good to urge teachers to call on low-achieving students more frequently if these students are generally asked to answer only simple factual questions or questions they cannot answer. Likewise, it is unproductive to seek to increase the amount of time that a teacher waits for a low-achieving student to respond, independent of a consideration of particular context. If the teacher asks a factual question, simply waiting and providing clues may be an unproductive use of classroom time (since the student either knows the answer or doesn't). However, in a situation involving judgment or analysis, more time to think and more teacher clues may facilitate a student response.

Teacher differences. Why do some teachers noticeably differentiate their behavior toward high- or low-achieving students, while others show only minimal differences in their behavior? Marshall and Weinstein (1984) have contended that teachers' beliefs about the nature of intelligence may influence their classroom behavior. Some teachers view intelligence as a fixed and relatively stable entity. Other teachers view intelligence as incremental, as a repertoire of skills and knowledge that can be increased.

Teachers who hold an entity view of intelligence seem more likely to place students in a stable hierarchy according to performance expectations and to treat these students differently on the basis of expectations. According to Marshall and Weinstein, low achievers in the classrooms of these teachers may use performance information to make detrimental social comparisons. However, the statements and curriculum assignments of teachers who hold an incremental view of intelligence communicate the belief that each student has the ability to improve regardless of current status, that individual differences in rates and modes of learning are normal, and that students can learn from those who have already acquired certain skills. Such distinctions about performance expectations seem particularly applicable to mathematics education.

Beliefs about mathematics. For a long time, people thought of mathematics as a way of representing true statements about nature. Mathematics was a means of knowing, a way of achieving certainty that relied on deductive proof from self-evident principles called axioms. Deductive reasoning, by its very nature, guarantees the truth of what is deduced, provided its assumptions are also true. As Kline (1980) notes, by utilizing this seemingly clear, infallible logic, ealy philosophers often thought that mathematicians inevitably produce irrefutable conclusions. It is now understood, however, that many forms of mathematics explore the implications of postulates that may or may not represent real-world events. Thus, it is quite possible for a given form of mathematics to be both logically *valid* and *inapplicable* to observed events. Unfortunately, many people continue to believe that mathematics produces irrefutable conclusions. And when writers want to provide examples of arguments that are both logical and valid, they often refer to mathematics. Unfortunately too, many teachers view mathematics largely as the production of correct answers. This is an unfortunate view that helps to account for the fact that mathematics teaching in elementary schools basically involves arithmetic computation and drill, with relatively little attention given to problem solving.

Worse, some teachers' goal seems to be to help students to produce *correct answers quickly* (see Thompson, 1985)—a view that is supported if not enhanced by textbooks. It also seems likely that teachers' conceptions of mathematics may, in combination with other beliefs and preferences, yield a base for predicting how teachers decide what is appropriate mathematics instruction for entire classes or for specific types of students. For example, some elementary school teachers who hold an entity view of intelligence, who tend to blame students for their failure, and who hold a "certainty" view of mathematics appear likely to stress mastery of basic concepts, to emphasize drill and practice, and perhaps to see instruction in problem solving as an enrichment topic—to be done if time allows. Moreover, students of teachers who held these beliefs and who perform poorly would also be likely to be assigned more practice, more drill. In such classes, the *solution becomes part of the problem*. Thus, students who may need both more structure and more practice receive only practice, making it difficult for them to understand the mathematics they are assigned.

Future research. Of the many issues for future research that are associated with expectations, we limit ourselves to commenting on only three concerns. First, it seems useful to conceptualize and to study teachers' mathematical expectations. It is important to identify teachers who have certain beliefs about mathematics and to determine how these beliefs interact with teachers' views of intelligence and of classroom learning strategies that are appropriate for particular students. Considering that we have a growing minority population, it is vital to learn how teachers' beliefs about ethnicity and what constitutes appropriate mathematics interact to form a basis for teachers' decisions about classroom instruction. All students can benefit from appropriate instruction in problem solving and estimation; however, some teachers may overreact to student or family characteristics, underestimate student potential, or provide students with only limited learning experiences.

Hodgkinson (1985) summarizes major educational consequences of recent demographic changes in the U.S. population. Among others, these changes include the appearance of: (1) more children who enter school from poverty households; (2) more children who enter from single-parent households; (3) more children from minority backgrounds; (4) a smaller percentage of children who have had Head Start and similar programs, even though more are eligible; (5) a larger number of children who were premature babies, leading perhaps to more learning difficulties in school; (6) more children whose parents were not married, now 12 of every 100 births; and (7) more children from teenage mothers. Thus, Hodgkinson concludes that the educational system must accommodate children who will be on average, poorer and more ethnically and linguistically diverse, and that by the year 2000 the United States will be a nation in which one of every three citizens will be nonwhite. Equally important is the understanding that minorities will cover a broader socioeconomic range than ever before, making simplistic understanding and treatment of minority needs even less useful. Considering such demographics, certain educators may lose sight of the fact that many minority students do exceedingly well in higher education, and there may be increased pressure for needless tracking of students in elementary schools.

A second concern reflects our present move toward a school curriculum that pays increasing attention to interdisciplinary conceptions of inquiry. Given such a stress, how should educators alter the behavior of students who have developed a passive orientation to learning and have come to accept the study of mathematics as a search for certainty? The work of Eder (1981), Good et al. (1987), and others suggests that students learn certain roles in school and that some of these roles are self-defeating. Researchers in mathematics education have reached similar conclusions. For example, Carpenter (1985) argues that certain students come to school with natural problem-solving abilities that are diminished by exposure to formula-algorithmic driven instruction. Educators must determine how to structure learning experiences for the vast number of students who have adopted a "certainty" view of mathematics. This is a complex question that deserves thoughtful research.

Some educators believe that the answer is to involve students in problem-solving inquiry in which they assume control of their own learning. We suspect that students who have limited understanding of mathematics, who search for "right" answers, who see the teacher as an authority figure, and who prefer to question the teacher rather than peers, may respond poorly to opportunities for problem solving. Carpenter is quite right in emphasizing that, for young children, some seemingly simple mathematics can represent "true" problems. Expecting too much too soon may be as inappropriate as expecting too little. We agree with Davis (1986), that

> the modern math curriculum should be extended to include such tasks and conditions as conjecturing what may be true or what may be useful; planning a strategy for attacking a problem or for attempting to invent (or discover) a proof; revising such plans; criticizing proofs, plans, or solutions; exploring potentially mathematical situations; seeking patterns, proving them, and using them; devising alternative representations; clarifying or explicating ambiguous or unclear information; generalizing; finding counterexamples. (p. 946).

However, one does not presently know how to advise a fourth-grade teacher concerning the approach advocated by Davis and others. We suspect that, unless the field of mathematics education proceeds from careful research and conceptualization, it will conclude twenty years or so from now what Powell, Farrar, and Cohen (1985) concluded about the activity curriculum: that the rhetoric, although motivating change in the short run, did not lead to a better understanding of how to achieve desired goals. Although it is by no means clear that calls for reform in mathematics education issued over the last ten years have had any effects on classroom practice, it seems likely that effective reform will have to be based on a firmer knowledge base than we now possess.

Third, and last, it seems likely that the teacher's role has been underestimated in recent discussions of elementary school mathematics curriculum improvement. Considering the growing interest in student variables and issues of student mediation, we believe that there is a tendency to overlook ways in which teachers may assist students to acquire problem-solving skills. After all, if students have learned to look for answers from teachers, it may be most useful—as a change strategy—for teachers to model how to search for and how to recognize multiple as opposed to single approaches to problem solving. To understand why teachers behave as they do, it appears necessary to understand also their beliefs about mathematics and classroom instruction. Simply put, we suspect that some teachers have more sophisticated views of mathematics and teaching and thus can more successfully integrate their knowledge of both content and processes into classroom settings. (Similar arguments have been made by Berliner, 1986 and Leinhardt, in press.)

One wonders what beliefs teacher education students have when they enter teacher education programs, what are the major dimensions of those belief systems that show variation, what are the factors associated with different belief systems. How and why do teachers' beliefs about mathematics change during preservice training, and how differentiated are those belief systems? Furthermore, it seems especially important to see how teachers' beliefs about mathematics interact with other belief systems (e.g., teachers' management systems or beliefs about student intelligence and learning) and to determine if belief systems influence teacher planning and other behavior. Needed also is research that examines how school factors (beliefs of principal, other teachers) and district policy affect teachers' belief systems and how these alter teaching practice. We believe that since teaching is an integrated experience, examination of one belief system alone (e.g., teachers' views of mathematics) will not yield models that accurately predict classroom behavior. However, information on a set of beliefs (mathematics, management, student learning) may enable researchers to identify cognitive factors that guide teacher planning and behavior in the classroom. To make such assessments, it will be necessary to build conceptual measurement inventories for collecting data about teachers' belief systems and to develop observational systems for assessing teacher and student behavior (related to the cognitive belief being explored). Similar attention might be given to understanding students' belief systems and how classroom factors affect those beliefs.

The Missouri Math Program

In the early 1970s, it became "fashionable" to believe that classroom teaching and schooling more broadly did *not* have important consequences for student learning. Fueled by the Coleman et al. (1966) and Jencks et al. (1972) analyses of school performance (which failed to include measures of classroom process), it was often concluded that no important variation existed between schools or classrooms. "If you've seen one teacher, you've seen them all," was the prevailing mood among social scientists, including many educators.

In sharp contrast, the decade of the 1970s also witnessed sharp expansion in the growth of observational research on teaching. Concern with what teachers actually do in the classroom led many researchers to focus on how teachers interacted with high- and low-achieving students. An incidental outcome of this research was the demonstration that teachers vary greatly across classrooms in their behavior, as well as in how they distribute their time and resources within classrooms. One of the authors and Doug Grouws became interested in testing the effects of teachers on students' mathematics learning (Good & Grouws, 1979). We wanted to assess the hypothesis that teachers *do* make a difference in student learning and to examine two broad questions: (1) What is the characterization of mathematics instruction in intermediate grades in elementary schools?, and (2) do certain forms of teacher behavior appear to have a more positive influence on student achievement than other patterns of teacher behavior?

Field observations. The research began with an initial study of over 100 third- and fourth-grade teachers who taught in the same school district and used the same textbook. To compare teachers for effectiveness, we used student performance on a standardized achievement test in mathematics as a way of estimating instructional progress.

Looking at test scores over a three-year period, we found that teachers varied considerably in their impact on students' progress, despite the fact that they were using the same textbook and in most cases were teaching students of comparable ability. Our initial data were a demonstration of an apparent *teacher effect*; some teachers produced more mathematics learning than did other teachers teaching in comparable settings.

Our next task was to observe teachers to find teaching strategies that were associated with high and low patterns of student progress. Accordingly, we decided to observe *stable* teachers (i.e., those who were consistently high or low across several consecutive years in their ability to produce student performance on standardized achievement tests). Interestingly, teachers who had differentiated effects on achievement (both high and low) used large-group teaching formats. Hence, format did *not* predict achievement. We found, however, that stable, high and low large-group teachers differed in their classroom behavior. That is, more and less effective teachers taught in different ways, and some of these differences in teaching behavior were consistent across the two groups of teachers. For example, teachers who obtained higher levels of student achievement, in contrast to teachers who obtained less achievement, provided students with a clearer focus of what was to be learned and provided greater amounts of developmental (process) feedback during instructional interactions.

The experimental study. Although we were pleased with our observational findings, we were aware that these were only correlational results and that they did not necessarily imply that these teacher behaviors *caused* student achievement. It could be that behaviors not studied in our observational research were more directly related to achievement. In particular, we wanted to see if we could instruct teachers to behave in ways consistent with the behavior of "effective" teachers and to determine what impact, if any, such behavior would have on student achievement. Because of the expense involved in field testing the program, we wanted it to be as comprehensive as possible. Thus, in addition to including the contrasts obtained in our earlier naturalistic studies, we wanted to test promising findings from other teacher effectiveness studies, as well as results from previous experimental mathematics studies (see Good, Grouws, & Ebmeier, 1983 for details).

We tested our program in 40 classrooms. About half of the classrooms were assigned to experimental conditions and the other half to control conditions. Several procedures were employed to ensure that the control group was motivated to pursue achievement gains in mathematics, so that a strong control for Hawthorne effects was built into the project (for details, see Good & Grouws, 1979). One major question was whether teachers would be willing to implement the program. On the basis of observers' records, it was found that most, but not all, of the experimental teachers implemented the program very well. Because experimental teachers *did* use the program, it was possible to determine how the experimental training and subsequent teaching activity influenced student achievement and attitudes.

Pre- and post-testing with the standardized achievement test indicated that, after 2-1/2 months of the program, the performance of students in experimental classrooms was considerably higher than that of those in control classrooms. It was also found that the experimental students' performance increase continued for at least some time following the treatment. Regular end-of-year testing by the public school system indicated that approximately three months after the program had ended the experimental students were still performing better than the control students. We also constructed a content test that more closely matched the material teachers were presenting. The results on this latter test also showed an advantage for experimental classes, although differences between control and experimental classrooms were not as large as they were on the standardized achievement test.

Research elsewhere has indicated that teachers have a favorable reaction to the program even when it is presented and discussed without the involvement of the developers (Andros & Freeman, 1981; Keziah, 1980). Also, in research at the junior high level, it appears that secondary teachers have implemented the program with positive impact on students' verbal problem-solving ability (Good & Grouws, 1981).

Our research on mathematics instruction, especially at the elementary school level, has convinced us that teachers *do* make a difference in student learning and that in-service teachers *can* be trained in such a way that student performance can be increased. The system of instruction that we see as important can be broadly characterized as *active teaching*. It is instructive to note that in our experimental work active teaching was an important difference between teachers who were getting good achievement gains and those who were getting lower-than-expected gains. Teachers whose students made higher gains were much more active in presenting concepts, explaining the meanings of those concepts, providing appropriate practice activities, and monitoring those activities prior to assigning seatwork. The fact that these teachers appeared to look for ways to check whether their presentations had been comprehended by students was particularly important. They assumed partial responsibility for student learning and appeared to be ready to reteach when necessary.

This difference in active teaching *across* classrooms is comparable to differences found *within* classrooms in teacher expectation research. That is, in the teacher expectation literature there is evidence that in some classrooms low-achieving students receive less active and less meaningful teaching than high-achieving students. In our effectiveness research in mathematics we have found that some teachers are less active and less effective in teaching the *entire* classroom. Active instructional efforts seem to be an important aspect of teaching that is related to achievement gain, at least in basic skill areas.

No simple answers. Data from the Missouri Math Program provide one model of mathematical instruction that appears to have merit for some teachers in some contexts. However, these data in no way imply that teachers should teach large groups of students or that teachers who choose to teach whole classes should use the model described by Good, Grouws, and Ebmeier (1983). As a case in point, recent work by Slavin and Karweit (1985) suggests that the instructional principles developed by research on the Missouri Mathematics Program can be effectively combined with other grouping and organizational plans. Needed are more models that might stimulate thoughts about other types of effective mathematics instruction.

As we stress shortly, research does not yield answers, but rather generates findings, concepts, and a degree of understanding. It should be clear that even in our carefully controlled study not all teachers implemented the program; hence, exposing teachers to materials and training associated with "Missouri Math" does not guarantee that teachers will fully implement the program. Continued classroom observational work and follow-up studies have provided and will continue to provide a more differentiated picture of what effects might be associated with the program. To illustrate, it now appears that "Missouri Math" generates more effects for some teacher and student combinations than for others (Ebmeier & Good, 1979). It also seems that the classroom organizational structure interacts with the effects of the instructional treatment (Good, Grouws, & Ebmeier, 1983).

It should also be stressed that there is *no* single system for presenting mathematics concepts effectively. For example, some of the control teachers in our studies have obtained high levels of student achievement using instructional systems that differed from those in the program we developed. More information about the classroom contexts and particular combinations of teachers and students that make the program more or less effective is clearly needed.

Research not only yields a set of findings, it also yields a way of looking at the problem and at concepts and criteria that can be applied in order to thoughtfully examine classroom instruction. Hence, in current work Good and Grouws are not advocating that teachers teach like the teachers in the Missouri Math program and implement behavior in a mechanical fashion. Rather, their efforts are to use the findings to stimulate teachers to discuss the criteria for guiding the quality of a development lesson. Given that the research suggests the need for a development phase of the lesson, how can teachers develop an image—a model—of what successful development looks like, and what are the various ways this model can be implemented in classrooms?

A major research effort now underway is to help teachers understand the criteria for a good development lesson. Teachers are being involved in discussing the criteria, in applying the criteria to videotapes of mathematics teaching, and in writing original development lessons and exchanging those lessons with other teachers. Hence, teachers are having the opportunity to build a theory of teaching—a set of beliefs and practices—that can be used to help students to understand the mathematics they study (Good and Grouws, 1987).

Future Research

Although the Missouri Math effort has many implications for needed research, we concentrate on three issues here. First, it is important to assess the relationship between teachers' presentations and students' attention. Teacher actions can either increase or decrease student attending and can thus influence whether student learning is appropriately or inappropriately directed. We suspect that the extreme presumption that students always listen to teacher talk and the alternative presumption that they always ignore teacher talk are both unsupportable. We need more information about the relationship between teacher behavior and student motivation. One important question is: What is the type of teacher behavior that increases the chance that students will actively seek to understand mathematics.

For example, Dillon (1982) argues that some educators presume a relationship between the cognitive level of questions that teachers ask and the cognitive level of student response. He presents convincing empirical data that the popular assertion, "ask a higher-level question, get a higher-level student response," should better be stated, "when teachers ask a higher-level question they get a any-level answer." However, he notes that when teachers' questions are distinguished from their statements, the relationship between teacher action and student response seems to be elevated for teacher statements and depressed for questions. The tendency, Dillon suggests, is for the following rule to apply: "Ask a high-level question, get a lower-level answer—but make even a low-level statement, and get a higher-level response" (p. 550). Hence, it seems reasonable to question the urging of some mathematics educators that teachers' role be limited to only that of raising questions and allowing students to discover mathematics. We suspect that most students will benefit from high-quality teacher statements about mathematics *content*.

Similarly, there seems to be a great deal of confusion about the role of student groups in learning basic skills, mathematical problem-solving, prosocial skills, and self-regulation strategies. We suspect that, at least in some forms, small-group work will make it more difficult for passive students to develop self-regulatory skills. Thus, educators should not presume that whole-class, individualized, or small-group learning has effects independent of how a program is implemented. Too many educators believe that whole-class instruction is necessarily dull and over-used, and that small-group instruction is full of potential and under-utilized. In essence, both of these strategies are problematic. More careful studies of the subtle communication that takes place in different models and a better understanding of how teachers utilize these formats and how students learn in them are needed.

Research has already suggested that the picture is not as simple as some reformers claim. For example, Bershon and Rohrkemper (1986) report that students' affective preferences for cooperative learning in small groups reflect different patterns for high- and low-ability students at various grade levels. Younger students with high math ability and older students with low math ability preferred to learn math in cooperative groups because they wanted to be able to receive and give help. One wonders if some students are developing subtle dependencies on other students.

While noting a number of positive things that were occurring during small-group discussion, Bershon and Rohrkemper report questionable process behaviors as well. For example, in student interviews, medium-ability students and older students did not recommend self-regulatory strategies and the advice recommended by low-ability students seems highly questionable. The possibility that peer communication of low expectancies and inappropriate strategies (as well as being helpful) has been empirically confirmed by Webb (1985), who illustrates that communication in student learning teams can have both negative and positive effects on student achievement. Again, we call for an examination of the reasons why teacher or peer communication may be more or less desirable for different types of mathematical learning rather than the glib assumption that teacher-directed or student-directed learning is superior, independent of the context in which instruction occurs.

Second, future research can assess some of the assumptions developed in the Missouri Math Program. For example, research is needed to determine if experimental teachers identify more student errors and can more readily understand those mistakes during the development stage than can control teachers who use different teaching techniques. It is equally important to determine whether students in experimental classrooms are more active thinkers during the development portion of lessons than are students in control classrooms (perhaps by asking students to do problems immediately after development). Similarly, future studies should identify the conditions under which student errors are developmentally helpful and lead to increased student effort to integrate material rather than debilitating and convincing students that they do not understand mathematics. When researchers, mathematics educators, and other educators examine their assumptions by stating and testing the specific ways in which student learning is influenced, the conditions under which certain teaching and learning strategies are useful will become clearer than they are at present.

Finally, a third topic that merits increased observational research is the comparison of expert and novice mathematics teachers. Berliner (1986) and Leinhardt (in press) build upon earlier teaching studies of more and less effective teachers by contrasting novice and expert teachers in order to extend knowledge of what effective instructors *do* in the classroom to include an understanding of *why* effective teachers behave as they do. Berliner notes that experts, in contrast to novice teachers, were able to categorize problems to be solved at a higher cognitive level; they could use higher-order systems of categorization to analyze problems; they were also extraordinarily fast and accurate at recognizing classroom events and were opportunistic planners who were able to change quickly when things were not moving at a productive pace.

Leinhardt (in press) reports on a comparison of a small group of expert teachers and novices and identifies several interesting differences. Among the differences she reports is the finding that expert teachers' explanations of content contained components that students already know; however, none of the novices' explanations included common phenomena that students know. Experts gave better explanations than novices and their explanations contained more critical features and fewer errors. Hence, experts know the content but also had ways of presenting that material in a meaningful fashion. Experts saw lessons as connected and helped students to see the connections.

The study of expert teachers is important, and *more* resources need to be made available for this important line of research. No doubt there are times at which young teachers will have to teach in different ways than do more experienced teachers (e.g., they may have to overplan, spontaneous moves may have more problematic results) as they have to learn to apply mathematics learning under the complex social conditions of classroom teaching. Hence, research on expert teachers needs to be balanced with continuing work on teachers of varying ability who teach in diverse situations.

Increasing the Role of Observational Research

We have discussed two examples of programmatic observational research to illustrate how research can yield concepts and new lines of inquiry. It seems important to argue at a more general level that observational research must be utilized if we are to understand classroom instruction and make appropriate recommendations for its improvement. Unfortunately, too often reformers seem willing to urge reforms on the basis of no evidence or survey evidence alone, without studying actual classroom practices. A good illustration of this point may be found in recent cross-cultural research.

In their report of the second IEA study of mathematics achievement, Crosswhite, Dossey, Swafford, McKnight, and Cooney (1985) note that despite the fact that U.S. teachers report emphasizing arithmetic at the eighth-grade level, comparative data indicate that the performance of U.S. eighth-grade students on arithmetic measures falls below that of eighth-grade students in many other countries. Thus, the survey suggests that if U.S. teachers *actually* place more emphasis on and spend *more time* teaching this material than do teachers in other countries, we have a problem: our response to this deficiency cannot simply be more time on mathematics, but rather, the quality of teaching must be improved.

Such findings have encouraged some educators to suggest that teachers spend too much time presenting information and that the way to improve teaching is to increase the percent of class time used for small-group instruction. There are at least two problems with this conclusion: it assumes that quality can be improved only by changing the form of teaching; and it overlooks alternative hypotheses, including the possibility that both the amount of teaching *and* the use of small-group instruction should be increased.

Some argue that teachers lecture too much and under-utilize grouping as an instructional format in mathematics. However, recent cross-national research that actually observed processes in classrooms sampled from suburban Minneapolis; Sendai, Japan; and Taipei, Taiwan (Stevenson, et al., 1987) raises some questions about the merits of this contention. Stevenson et al. (1987) report that U.S. teachers spend considerably *less* time imparting information than do Chinese and Japanese teachers. Furthermore, when U.S. teachers do talk, they are more likely to give *directions* than to discuss content. In contrast, Japanese and Chinese teachers are more likely to provide content when they talk. Clearly, teacher "talk" or "lecture" are vague concepts that may describe various types of teaching. Still, these data do not support the simple suggestion that U.S. teachers talk too much.

Classroom observation may change not only inferences made about the survey data but the validity of the data as well. For example, the data of Stevenson et al. indicate that U.S. teachers spend less time teaching mathematics than do teachers in the other two countries. Chinese and Japanese students spend twice as much time studying mathematics as U.S. pupils. Hence, the time spent on mathematics as well as the quality of teaching may be associated with poorer student performance in U.S. classrooms. Indeed, there may be too little effective teaching of arithmetic concepts and too little effective teaching of problem-solving skills in U.S. schools. These data suggest that another problem may exist as well, assuming that time allocations represent some normative dimension. Perhaps less time is allocated to mathematics because United States citizens believe that mathematics instruction is less important. If so, changes in classroom teaching alone may be insufficient for increasing the mathematics performance of average and less able students.

What U.S. teachers and students spend time on may be a major part of the problem (as well as the amount of time allocated and the quality of mathematics teaching). For example, the data presented by Crosswhite et al. (1985) suggest that too much time in U.S. classrooms is spent on review and grading papers and too little time is spent developing *new content*. Unfortunately, observational data were not collected by Stevenson et al. (1987) to test this latter hypothesis, but it certainly seems viable. To spend time on review and noninstructional issues would certainly seem to interfere with developing new content.

Our comments are not intended to discourage the use of small groups in mathematics instruction, for we suspect that both more small-group instruction and more teaching should occur in U.S. classrooms. For example, both the survey data of Crosswhite et al. and the observational data of Stevenson et al. indicate that U.S. teachers use an inordinate amount of seatwork and underscore the usefulness of increasing the use of student groups in various areas of mathematics instruction.

However, we do not assume that simply increasing the amount of small-group instruction would necessarily make the discussion of mathematics more meaningful nor practice more effective. Although we do not see small-group instruction as a panacea, we do see it as a viable instructional format when it follows careful principles of organization and implementation (see Slavin, 1983). We suggest that small-group instruction can be used to achieve a variety of goals: effective practice, the opportunity to learn prosocial skills, and the value of taking different approaches to solving problems. However, if the goal is to teach autonomy or individual persistence, different models may be better. All of our speculations await observation data. Research on small-group instruction in mathematics has illustrated only that the motivational components of small-group work can be powerful determinants of learning basic skills (Slavin, 1983). Yet to be demonstrated is whether small-group instruction has important consequences for how students think about mathematics content.

Indeed, work by Noddings (1985) suggests that thoughtful whole-class instruction can produce as much discussion and positive problem-solving behavior as does small-group instruction. Again, as we have argued before, the quality of mathematics instruction is more important than the form, and observation is an important way to learn about the quality that lies behind the image painted by survey data.

Limitations of Observational Research

The major thrust of our argument is that increased effort and funding should be directed towards observational research on mathematics education. It is possible to misconstrue the thrust of that argument, however. Let us examine some of the limitations of observational research, as sometimes practiced or conceived by others, to see what we are *not* arguing.

Policy Recommendations

Some scholars have urged that observational research on teaching be pursued because it generates findings that have immediate and obvious implications for educational policy. And, in pursuit of this vision, some research has appeared during the past 15 years that reports covariations between classroom events and pupil learning in the hope that these findings can guide immediate and universal efforts to improve the subsequent conduct of teaching. Moreover, examples may be cited to show that policymakers have been influenced by the findings of classroom research. To illustrate, influential studies have shown that pupil learning is greater when teachers spend more time-on-task in their classrooms, and this finding has affected policymakers who argue that teachers should spend more minutes during the school day, or hours during the school year, on subject-matter instruction. For a good illustration of this misuse see the discussion in *A Nation at Risk.*

Unfortunately, there are things wrong with this simplistic vision. For one, the findings from which policy implications are argued are rarely explored for the full range of classroom events to which we would like to make policy recommendations. (Let us assume that increasing time-on-task will increase pupil learning in the "typical" classroom which presently spends but little effort on subject-matter instruction. What would be the effect of demanding an increase in time-on-task for classrooms that already devote a good deal of effort to subject-matter instruction? Is it not possible that the effect is curvilinear and that pupils may become bored or fatigued if subjected to too much time-on-task?)

For a second reason, findings from classroom research are always generated in a specific observational or experimental context whereas policy recommendations have their impact in other settings. Unfortunately, many policy recommendations have unintended effects when implemented in the latter. (Again to illustrate, reports have appeared suggesting that when teachers are told that they must increase pupil learning and that the best way to do this is to increase time-on-task they will demand that *their* classroom lessons take precedence over extra-curricular experiences, school assemblies, and even the lessons of other teachers!)

But above all, it is an error to assume that findings from classroom research can ever stand alone. Findings from this form of research are like all research findings; relatively meaningless when considered in isolation, they gain their meaning and significance when they are interpreted through theory. Thus, with Kerlinger (1977) and Biddle and Anderson (1986), we insist that the fundamental task of conducting classroom research is to generate and test *plausible theory* concerning teaching. When relatively primitive, such theories consist of terms that represent classes of observable events and propositions that summarize observed relationships among them. But better, and most persuasive, theories include explanations for observed events, and it is the fundamental task of classroom research to generate and test the latter. And should a particular program of classroom research generate plausible theories about teaching that are well argued and well supported with evidence, it should not surprise us that presently those theories will begin to affect the decisions of policymakers in education.

Inherent Limitations

Observational research in classrooms is also subject to inherent limitations, and it is wise to bear these in mind when arguing for the advantages of this form of research. For one thing, the collection of observational data in classrooms is an expensive, labor-intensive business, and this tends to limit the size and representativeness of samples of classrooms, teachers, pupils, and classroom events studied. This problem is addressed in several ways by researchers. When interested in studying longer sequences of teaching events (i.e., those that evolve over a series of lessons) researchers tend to study but one or two classrooms (e.g., Smith & Geoffrey, 1968; Carew & Lightfoot, 1979). But even when their interest is focused on teachers' behaviors alone or on short sequences of teacher-pupil interaction, researchers still find it difficult to study more than a small sample of classrooms, and unfortunately that sample is nearly always drawn for convenience rather than randomly. Thus, it should not surprise us to learn that much of past classroom research has been conducted in contexts that feature white, middle-class pupils and willing teachers. One makes generalizations from such limited and biased samples with hesitancy.

Let us take the problem above seriously and presume that significant differences occur among classrooms depending on pupil population, teacher personality characteristics, and cultural context. One way to investigate such phenomena is to conduct comparative research in which we juxtapose classrooms from various contexts to see in what ways they differ. To illustrate, one might compare mathematics lessons from middle-class schools with those from schools serving large numbers of students from low-income homes. Or one might compare mathematics lessons taught at various grade levels. Or one might examine mathematics instruction in schools that used a lock-step curriculum with those that encouraged laissez faire instruction, or comparisons might be made in those schools that make extended use of computers and calculators and those that do not. Unfortunately, such studies are rare.

In part, these problems may be traced to lack of vision on the part of researchers and funding agencies, but above all research on teaching has been throttled by lack of finance. Unfortunately, it takes a good deal of money to collect, transcribe, code, and analyze data from observations of teaching, and one conducts this type of research only when funds are available to finance the effort. Serious research on teaching did not appear until the 1960s when federal funds were at least available to support educational research, and with the decline of that support in the Reagan years research on teaching has been severely curtailed. Unless federal or alternative sources of finance are made available, observational research on teaching will shortly disappear.

It should also be pointed out that the classroom is not the sole stimulus environment to which the pupil is exposed, and it is quite possible that other environments will also affect achievement in mathematics or other subjects. To illustrate, let us consider the widespread concern for differential achievement in mathematics by high-school-age boys and girls. As far as one can presently tell, this phenomenon is associated with differential willingness of boys and girls to enroll in advanced mathematics courses. (Most studies indicate but minor differences in mathematics achievement between boys and girls as long as both sexes are enrolled in the same courses—for example, see Fennema & Sherman, 1977—but girls seem less willing than boys to enroll in higher-level, elective, mathematics courses, and their accomplishments begin to lag behind those of boys once they are allowed to drop out of the chase.) At this writing, it seems unlikely that girls' unwillingness to enroll in higher-level mathematics courses is related only to experiences they have suffered in their classrooms. Instead, one suspects that the problem lies in an association between sex-role stereotypes and mathematics—stereotypes that appear first in the home but are also bolstered through informal interaction among boys and girls in out-of-classroom contexts as well as in classroom exchanges.

This discussion of external influences on students' beliefs and performance is not intended to suggest that the study of gender issues in mathematics classroom is unimportant. Indeed, much of value has been learned about classroom factors that do unduly restrict the mathematics performance and aspirations of female students (e.g. see Fennema and Meyer, in press). However, the example is to illustrate that there are limits to what observational research can do and that any classroom issue—teacher expectations, student beliefs—are influenced to some extent by external factors.

Research on Other Topics

These inherent limitations suggest that, however important one may consider research on classroom teaching, one should never forget that it is but part of the larger corpus of educational research. Through observational research one can generate unique and utterly vital findings about mathematics instruction, but one must supplement those findings with information from other types of study.

To illustrate this point, Romberg and Carpenter (1986) provide an insightful summary of recent research on the learning of mathematics concepts among primary pupils. Since the focus of this latter research is upon cognitive processes in pupils that cannot be observed directly, studies within its ambience are unlikely to look much like observational studies of teaching. Instead, the typical study of mathematics learning involves interviews with pupils or the manipulation of pupil concepts in experiments. There is nothing wrong with such research designs; indeed, they would appear extremely useful, and the knowledge they help to generate is clearly necessary if we are to understand why mathematics instruction produces the effects in pupils we observe. Thus, research on the learning of mathematics instruction is a necessary complement to research on the teaching of mathematics.

But the need does not stop with research on pupil learning. In order to understand why different patterns of mathematics instruction appear, we also need research on community beliefs about mathematics, school climates, and curricular impact, and if we are to take effective actions to improve mathematics instruction we also need to know how teachers think about their teaching and how those thoughts are translated into classroom strategies. In short, research on teaching cannot stand alone but must be integrated within a larger research effort. Our arguments for increased efforts and funds for research on the teaching of mathematics should be understood to include arguments also for increased efforts and funds for other, related research efforts.

REFERENCES

Andros, K., & Freeman, D. (1981). *The effects of three kinds of feedback on math teaching performance*. Paper presented at the annual meeting of the American Educational Research Association, Los Angeles.

Artley, A. (1981). Individual differences and reading instruction. *Elementary School Journal, 82*, 143-151.

Berliner, D. (1986). In pursuit of the expert pedagogue. *Educational* Researcher, 15, 5-13.

Bershon, B., & Rohrkemper, R. (1986). *Elementary students' perceptions and management of classroom resources*. Paper presented at the annual meeting of the American Educational Research Association, San Francisco.

Biddle, B. J., & Anderson, D. S. (1986). Theory, methods, knowledge, and research on teaching. In M. C. Wittrock (Ed.), *Handbook of* research on teaching (3rd ed., pp. 230-252). New York: Macmillan.

Brophy, J. (1981). Teacher praise: A functional analysis. *Review of* Educational Research, 51, 5-32.

Brophy, J. (1983). Conceptualizing student motivation. *Educational* Psychologist, 18, 200-215.

Brophy, J., & Good, T. (1970). Teachers' communications of differential expectations for children's classroom performance: Some behavioral data. *Journal of Educational Psychology, 61*, 356-374.

Brophy, J., & Good, T. (1974). *Teacher-student relationships: Causes and consequences*. New York: Holt, Rinehart and Winston.

Carew, J., & Lightfoot, S. (1979). *Beyond bias*. Cambridge, MA: Harvard University Press.

Carpenter, T. (1985). Learning to add and subtract: An exercise in problem solving. In E. Silver (Ed.), *Teaching and learning mathematical problem solving: Multiple research perspectives* (pp. 17-40). Hillsdale, NJ: Lawrence Erlbaum Associates.

Chall, J. (1967). *Learning to read: The great debate*. New York: McGraw-Hill.

Claiborn, W. (1969). Expectancy effects in the classroom: A failure to replicate. *Journal of Educational Psychology, 60*, 377-383.

Coleman, J., Campbell, E., Hobson, C., McPartland, J., Mood, A., Weinfield, F., & York, R. (1966). *Equality of educational opportunity*. Washington, DC: U.S. Government Printing Office.

Commission on Reading. (1985). *Becoming a Nation of Readers*. Washington, DC: The National Academy of Education.

Cooper, H., & Good, T. (1983). *Pygmalion grows up: Studies in the expectation communication process*. New York: Longman. Crosswhite, F., Dossey, J., Swafford, J., McKnight, C., & Cooney, T. (1985). *Second international mathematics study: Summary reports for the United States*. Champaign, IL: Stipes.

Davis, R. (1986). What's new in math ed?: A review of E. Silver (Ed.), Teaching and learning mathematical problem solving: Multiple research perspectives. *Contemporary Psychology, 31*, 945-947.

Dillon, J. (1982). Cognitive correspondence between question, statement, and response. *American Educational Research Journal, 19*, 540-551.

Doyle, W. (1983). Academic work. *Review of Educational Research, 53*, 159-199.

Dunkin, M., & Biddle, B. (1974). *The study of teaching*. New York: Holt, Rinehart and Winston.

Dusek, J. (Ed.). (1986). *Teacher expectancies*. Hillsdale, NJ: Lawrence Erlbaum Associates.

Ebmeier, H., & Good, T. (1979). The effects of instructing teachers about good teaching on the mathematics achievement of fourth-grade students. *American Educational Research Journal, 16*, 1-16.

Eder, D. (1981). Ability grouping as a self-fulfilling prophecy: A microanalysis of teacher-student interaction. *Sociology of* Education, *54*, 151-161.

Eli Lilly and Company. (1986). *Lilly: Annual Report, 1986.* Lilly Corporate Center, Indianapolis, IN.

Evertson, C. & Green, J. (1986). Observation as inquiry and method. In M. C. Wittrock (Ed.), *Handbook of Research on Teaching* (3rd ed., pp. 162-213). New York: Macmillan.

Fennema, E. & Meyer, M. (In press). Girls, boys, and mathematics. In T. Post (Ed.), *Research-based methods for teachers of elementary and middle school mathematics.* Newton, MA: Allyn & Bacon.

Fennema, E., & Sherman, J. (1977). Sex-related differences in mathematics achievement, spatial visualization, and affective factors. *American Educational Research Journal, 14*, 51-72.

Flesch, R. (1955). *Why Johnny can't read and what you can do about it.* New York: Harper & Brothers.

Futrell, M. (1986). Restructuring teaching: A call for research. *Educational Researcher, 15*, 5-8.

Good, T. (1981). *Listening to students talk about classrooms.* Paper presented at the annual meeting of the American Educational Research Association, Los Angeles.

Good, T. (1986). *The role of teaching research in school reform, classroom practice, and teacher education: Past, present, and future.* Paper presented at the Chancellor's Invitational Conference on Education. Chapel Hill: University of North Carolina.

Good, T., & Brophy, J. (1986a). School effects. In M. Wittrock (Ed.), *Handbook of research on teaching* (3rd ed., pp. 570-602). New York: Macmillan.

Good, T., & Brophy, J. (1986b). *Educational psychology: A realistic approach* (3rd ed.). New York: Longman.

Good, T., & Brophy, J. (1987). *Looking in classrooms* (4th ed.). New York: Harper & Row.

Good, T., & Grouws, D. (1979). The Missouri Mathematics Effectiveness Project: An experimental study in fourth-grade classrooms. *Journal of Educational Psychology, 71*, 355-362.

Good, T., & Grouws, D. (1981). *Experimental research in secondary mathematics classrooms: Working with teachers* (Final Report of the National Institute of Education Grant NIE-G-79-0103). Columbia, MO: University of Missouri-Columbia.

Good, T., & Grouws, D. (1987, June). Developing inservice training programs for increasing understanding of mathematical ideas. *Phi Delta Kappan*, 778-783.

Good, T., Grouws, D., & Ebmeier, H. (1983). *Active mathematics teaching.* New York: Longman.

Good, T., Slavings, R., Harel, K., & Emerson, H. (1987). Student passivity: A study of student question asking in K-12 classrooms, Sociology of Education, *60*, 181-199.

Good, T., & Weinstein, R. (1986a). Schools make a difference: Evidence, criticisms, and new directions. *American Psychologist, 41*, 1090-1097.

Good, T., & Weinstein, R. (1986b). Teacher expectations: A framework for exploring classrooms. In K. Zumwalt (Ed.), *Improving teaching: 1986 ASCD yearbook* (pp. 63-85). Alexandria, VA: Association for Supervision and Curriculum Development.

Hodgkinson, H. (1985, June). *All one system: Demographics of education, kindergarten through graduate school.* Washington, DC: Institute for Educational Leadership, Inc.

Jencks, C., Smith, M., Acland, H., Bane, M., Cohen, D., Gintis, H., Heyns, B., & Michelson, S. (1972). *Inequality: A reassessment of the effect of family and schooling in America.* New York: Basic Books.

Kerlinger, F. N. (1977). The influence of research on education practice. *Educational Researcher, 6*(8), 5-12.

Keziah, R. (1980). Implementing instructional behaviors that make a difference. *Centroid* (North Carolina Council of Teachers of Mathematics), 6, *2-4.*

Kline, M. *(1980). Mathematics: The loss of certainty.* New York: Oxford University Press.

Leinhardt, G. (in press). Math lessons: A contrast of novice and expert competence. *Educational Psychologist.*

Marshall, H., & Weinstein, R. (1984). *Classrooms where students perceive high and low amounts of differential teacher treatment.* Paper presented at the annual meeting of the American Educational Research Association, New Orleans.

Mittman, A. (1985). Teachers' differential behavior toward higher- and lower-achieving students and its relationship with selected teacher characteristics. *Journal of Educational Psychology, 35,* 149-161.

National Commission on Excellence in Education. (1983). *A nation at risk: The imperative for educational reform.* Washington, DC: National Institute of Education.

Noddings, N. (1985). Small groups as a setting for research on mathematical problem solving. In E. Silver (Ed.), *Teaching and learning mathematical problem solving: Multiple research perspectives* (pp. 345-359). Hillsdale, NJ: Lawrence Erlbaum Associates.

Powell, A., Farrar, E., & Cohen, B. (1985). *The shopping mall high school: Winners and losers in the educational marketplace.* Boston: Houghton-Mifflin.

Romberg, T.A. & Carpenter, T. P. (1986). Research on teaching and learning mathematics: Two disciplines of scientific inquiry. In M. C. Wittrock (Ed.), *Handbook of research on teaching* (3rd ed., pp. 850-873). New York: Macmillan.

Rosenthal, R., & Jacobson, L. (1968). *Pygmalion in the classroom: Teacher expectation and pupils' intellectual development.* New York: Holt, Rinehart and Winston.

Slavin, R. (1983). *Cooperative learning.* New York: Longman.

Slavin, R., & Karweit, N. (1985). Effects of whole class, ability grouped, and individualized instruction on mathematics achievement. *American Educational Research Journal, 22,* 351-368.

Smith, L., & Geoffrey, W. (1968). *The complexities of an urban classroom.* New York: Holt, Rinehart and Winston.

Sternberg, R. (1985, November). Teaching critical thinking, Part I: Are we making critical mistakes? *Phi Delta Kappan,* 194-198.

Stevenson, H. W., Stigler, J. W., Lucker, G. W., Lee, S., Hsu, C. C., & Kitamura, S. (1987). Classroom behavior and achievement of Japanese, Chinese, and American children. In R. Glaser (Ed.), *Advances in instructional psychology, Vol. 3* (pp. 153-204). Hillsdale, NJ: Lawrence Erlbaum Associates.

Thompson, A. G. (1985). Teachers' conceptions of mathematics and the teaching of problem solving. In E. A. Silver (Ed.), *Teaching and learning mathematical problem solving: Multiple research perspectives,* (pp. 281-294). Hillsdale, NJ: Lawrence Erlbaum Associates.

Webb, N. (1985). Student interaction and learning in small groups: A research summary. In R. Slavin, S. Sharan, S. Kagan, R. Hertz-Lazarowitz, C. Webb, & R. Schmuck (Eds.), *Learning to cooperate, cooperating to learn* (pp. 147-172). New York: Plenum Press.

NOTE

[1]We acknowledge the support provided by the Center for Research in Social Behavior, University of Missouri-Columbia. We especially thank Cathy Luebbering and Diane Chappel for typing the manuscript.

From Fragmentation to Synthesis: An Integrated Approach to Research on the Teaching of Mathematics

Celia Hoyles

University of London Institute of Education
20 Bedford Way, LONDON WC1H OAL

This paper reviews some major research themes of relevance to mathematics teaching; that is, themes concerned primarily with teachers, with pupils and with classroom practice. A theoretical model for guiding future research effort is suggested. This model is based on a synthesis of (a) commitment to pupil activity and autonomy with (b) notions of pedagogic guidance and feedback drawn from analysis of the logic and meaning of some mathematical content and the pupil conceptions of this content. The central recommendations of the paper are: (1) the need to preserve the complexity of the research problematic; that is, to embrace a three-focus perspective on teacher, pupil and mathematics within a classroom environment; (2) the need to work with teachers as research collaborators; and (3) the need to acknowledge that teacher and pupil behavior in the context of a classroom cannot be understood through analysis of mathematical content or its related psychological complexity alone. The recontextualisation of mathematics in order for it to be socially enacted in schools and the constraints of classroom relationships must also be taken into account. Computer use as a special learning environment within the theoretical model is proposed as a catalyst with the potential to provoke new perspectives on school mathematics and how it should be taught.

"Chris Smith is a really good mathematics teacher!" What picture does this statement about Chris Smith conjure up for you? Does your picture change if you are told that the comment above was made by a school mathematics inspector? a pupil? another teacher? How does your picture change if you now find out that Chris is a woman (or a man) or that the comment was made by a male or by a female? What do we actually mean when we say someone is a good mathematics teacher? Is there something different in being a good mathematics teacher as opposed to being a good history teacher or even just a good teacher? This question was asked in 1983 by a small research project, the Mathematics Teaching Project (Hoyles, Armstrong, Scott-Hodgetts & Taylor, 1984). This project has conducted case studies of six teachers in three mixed London comprehensive schools, with two teachers in each school. The teachers were known for their enthusiastic and competent approach to mathematics teaching. Two fourth year (14-year-old) classes were studied for each teacher, one class of high attainment in mathematics and one of middle/low attainment. Some of the findings of this project will be referred to in this paper alongside work in other countries and from other perspectives.

RESEARCH ON TEACHING

It was noted in the NCTM president's report (Crosswhite, 1986) that the theme derived for the 1986 NCTM annual conference was "Better Teaching, Better Mathematics." Crosswhite also commented that the late Ed Begle had observed that we have learned a lot about "teaching better mathematics but not much about teaching mathematics better" (p.54). Crosswhite went on to suggest that school mathematics has suffered over the years from a range of false dichotomies, for example, "the old and the new in mathematics, the concrete and the abstract." Similar dichotomies are apparent in discussion of teaching mathematics, especially in the divide between proponents of exploratory discovery-based, inductive learning and proponents of more highly structured, didactic instruction. Sharply contrasting educational philosophies are evident in the approaches taken by these different groups (cf. Shulman & Keislar, 1966). These philosophies are concisely summarized by Lepper (1985).

> To the adherents of inductive, exploratory, Piagetian learning, the virtues of self-directed discovery arising from a sustained involvement in an educationally rich simulation . . . are substantial. Not only will such an experience heighten students' motivation; it will also produce deeper, more lasting, more meaningful learning (e.g., Brown, 1984; Bruner, 1961, 1966; diSessa, 1982; Lawler, 1981, 1982; Papert, 1972, 1980). Equally clear, however, is the challenge posed by supporters of a more explicitly didactic approach (e.g., Becker, 1971; Englemann, 1968; Skinner, 1968). They propose that the same knowledge, strategies, and powerful ideas children acquire inductively could be taught more efficiently and effectively with a more directive and structured approach to instruction. The tension and the interesting research issues involve the potential trade-off between "mere efficiency" of possibly superficial learning and the "depth" of learning presumed to arise when students themselves discover, induce, and interact with intellectually powerful ideas. (p. 13)

It is my belief that we must try to capitalise on the potential advantages of self-directed learning whilst ensuring that pupils come to grips with the intricacies of the subject matter itself. But first let us examine the background situation in more detail.

In a paper presented to the Special interest Group on Research in Mathematics Education at the annual meeting of the AERA, April 1986, Jere Brophy gave a wide-ranging summary of research on teaching, mainly in the U.S., which, though not conducted with a specific focus on mathematics, had relevance, in his view, to mathematics education. The paper reviewed process-outcome research, which links teacher behavior to student achievement, and described instructional strategies which were said to comprise effective teaching. Brophy drew attention to the review by Brophy and Good (1986) as follows:

> . . . the academic learning time that is most powerfully associated with achievement gains is not merely "time on task", or even "time on appropriate tasks", but time spent being *actively taught* (my emphasis) or a least supervised by the teacher. Greater achievement gains are seen in classes that include frequent lessons (whole-class or small group, depending on grade level and subject matter) in which the teacher presents information and develops concepts through lecture and demonstration, elaborates this information in the feedback given following responses to recitation or discussion questions, prepares the students for follow-up assignments by giving instructions and working through practice examples, monitors progress on those assignments after releasing the students to work on them independently, and follows up with appropriate feedback and reteaching when necessary. The teachers in such classes carry the content to their students personally rather than leaving it to the curriculum materials to do so, although they usually convey information in brief presentations followed by recitation or application opportunities. (Brophy, 1986, p.8)

Brophy went on to suggest that:

> process-outcome research has yielded information about the procedural specifics or qualitative nuances involved in implementing active instruction. Topics include the structuring and sequencing of the content to be presented, the clarity and enthusiasm of the presentation, the difficulty levels and cognitive levels of the questions asked of the students, which students teachers call on and how response opportunities are allocated, feedback reactions to successful and unsuccessful student responses, methods of attempting to elicit an improved response, praise and criticism, use of student ideas, and seatwork management. (Brophy, 1986, p.8)

Brophy's paper was criticized from a constructivist viewpoint by Jere Confrey (Confrey, 1986). Confrey attempted to highlight the fundamental differences in educational philosophy between her model of mathematics education and that put forward by those concerned with active mathematics teaching by creating a Socratic dialogue between Socrates and a researcher on teaching. The culmination of the dialogue was the apparent contradiction between promotion of intrinsic motivation by providing opportunities for pupil generated answers and effective teaching involving structured presentation and explanation of academic subject matter. Cobb (1986), in a similar way to Confrey, argued that "a rigid adherence to the transmission view of instruction . . . is incompatible with two goals of mathematics instruction valued by the constructivist—the construction of increasingly powerful conceptual structures and the development of intellectual autonomy . . ." (Cobb, 1986, p. 19).

One of the aims of this paper is to suggest that a way forward is first to acknowledge the essentially dialectic relationship between effective teaching and constructivism and then to find ways to build upon their complementarity.

Teaching Mathematics and not just Teaching

There have been many studies that have looked at the relative effectiveness of different teaching approaches across several subject areas. One such study in U.K. was the ORACLE research programme (Galton, Simon, & Croll, 1980). This research focused on teachers' strategies and tactics and on the interaction processes in the classroom. It reported that, "forms of classroom organization and consequent teacher-pupil and pupil-pupil interaction patterns are important in terms of educational outcomes" (p. 170). Six different teaching styles were identified in the ORACLE research (including for example class enquirers, group instructors, individual monitors), each derived from analysis of interactional data obtained through systematic observation. It was then noted that some of these teaching styles were more effective than others in terms of the pupil outcomes measured in the study. The Oracle evidence pointed to the widespread reliance on individualization of the teaching/learning process in U.K. primary schools. It was the most popular method used by teachers across all styles and, in the case of one style, "individual monitors," almost the only method used. The promotion of discovery or enquiry, the stimulation of pupils to thought and creativity, seemed, according to the researchers, to be minimized in this situation, since the dominant mode of teacher-pupil interaction was inevitably didactic given normal classroom conditions. It was therefore suggested that probing and stimulating questions and statements of the "higher order" kind were maximized by teachers of whatever style in a *class-teaching situation*. In addition to a study of teachers, the ORACLE research collected data derived from observation of a sample of eight pupils (representative in terms of achievement and sex) in each classroom. This work led the researchers to propose that some consistent relationships could be discerned between teacher style and pupil mode of working.

One major omission in the ORACLE research was any explicit consideration or analysis of the *content* taught. This is not an isolated phenomenon and many other projects have similarly sought to investigate the broad question of what makes for effective teaching or what styles of teaching are most productive. It is interesting to note that many researchers of teaching have in fact chosen to study classrooms where mathematics teaching is taking place—even though they do not address the *nature* of the mathematical knowledge being taught. One is then led to ask whether this preference arises from a view of mathematics as something which is invariant, as a paradigm of knowledge demonstrating certainty and universality. If this is the case one must then ask how such a view might influence subsequent definitions of effective teaching. These general studies have not, in my view, proved particularly fruitful. It would seem unlikely that there are many techniques of teaching that are necessarily desirable in all situations, given that different situations involve not only different personnel but also different purposes. In the rationale for the Mathematics Teaching Project it was stated that "good mathematics teachers require certain qualities which vary, at least in rank, from those of teachers in general" and that research on teacher effectiveness needs to pay attention to effectiveness within a specific subject area as opposed to consideration of a global concept. The point to be made is neatly put by Barrow (1986):

> One might wish to conduct research into the properties of a good car. But it is likely that, for practical purposes, knowledge of what is common to good cars will need to be supplemented by further research into the more specific questions of what makes a good racing car, what makes a good family saloon what makes a good farm vehicle, etc. Just as research into the former will require an adequate conceptualization of "a good car" before it can look into questions of design and structure, so the latter will require adequate conceptualization of the more specific notions of a good racing car, etc. So far as research into classrooms and teaching goes, this more specific conceptualization will need to take the form of conceptualizing the particular content in question. (p. 266)

There is now some empirical evidence that supports the contention that subject matter plays an important role in the thinking of teachers (even when maintaining discipline) and that information about the pupil is frequently used with reference to the content of the subject matter (Clark and Peterson 1986). After a study focusing on teacher explanations in mathematics classrooms, Bromme and Juhl (1984) concluded that:

> . . . the prevalence of explanations referring to tasks and specific areas of subject-matter as opposed to explanatory concepts extending across tasks, indicates that the respective specific elements of subject matter content within the perception and explanation of psychological facts (such as pupil understanding) should be increasingly taken into account in research on teachers' cognition. (p. 8)

It therefore seems apparent that researchers into teaching should look at what is being taught as well as by whom and to whom. They might expect that what is interesting and fruitful in their investigations will differ with different teachers, children, and subject matter. Each one of these three poles is itself complex. Mathematics, for example, is not a unitary subject and its critical features vary across topics. The teacher and pupil poles can also be researched through different lenses as indicated in the following sections.

The Pupil Perspective

Learning is dependent upon the situations and circumstances in which it is engendered and the feelings these situations provoke in pupils. Many studies of classrooms have neglected to take into account pupils' expectations, perceptions of school mathematics, and interpretations of classroom events. Yet every teacher knows that these have a profound influence on classroom behavior and achievement. We know that some pupils are either anxious about mathematics, alienated from it, or simply bored by it. They do not expect mathematics to demand thought or creativity but rather perceive it as an "obstacle" to be overcome or avoided. Problems are classified as hard or easy and solutions as right or wrong in ways often perceived by the pupil to be completely beyond personal control or comprehension. Reactions to mathematics which are typical of this group of pupils are illustrated in the following extracts from pupil interviews:

"... [It's] always the same; we never do anything different and then we never learn anything ... we don't learn nothing and then next year because we haven't learnt nothing, we will have to do it again ..."

"Well there is maths all this year. I just cannot do it. I cannot remember what it was even but it should all be easy. I just find it hard and it is all the easy stuff ... I keep trying and trying and nothing comes out. I feel so tight inside I want to explode ... when I am sitting there I know I will not be able to do it ..." (Hoyles, 1982, pp. 362-3)

The study of the pupil perspective in the Mathematics Teaching Project had the following design:

1. Elicitation, from all the pupils in the twelve experimental classes, of written descriptions of an "ideal" mathematics teacher.

2. Extraction from these descriptions of up to ten of the most frequently mentioned factors.

3. Marking of the above factors by the use of a paired-comparison test completed by all the pupils.

4. Grading of the teachers by each of their pupils on a numeric scale for each of the factors.

5. Individual interview of each pupil in order to obtain a description of classroom events chosen by the pupil as practical manifestations of one or more of the positive characteristics which they had attributed to their teacher. (Hoyles et al. 1984, p. 30).

This methodology allowed a comparison of the same teacher from the perspective of two different pupil groups who were of the same age but differentiated in terms of mathematical attainment. Analysis is still ongoing. Some results of the pilot study (reported in Hoyles, Armstrong, Scott-Hodgetts & Taylor, 1985) with respect to one teacher and one high attainment class are given in Tables 1 to 3 for the purpose of illustration. In addition to this data the pupil interviews provide additional qualitative data concerning the kinds of episodic judgments lying behind the pupil ratings given in Table 3.

The focus of the pupil perspective in the Mathematics Teaching Project was essentially the pupil expectations of their mathematics teacher. The project was not set up to consider the relationship of pupil perspective with any particular mathematical content, and indeed this did not arise in the pupil interviews. Although we wished to take into account the content to

Table 1

Desirable Characteristics in a Mathematics Teacher

A good mathematics teacher:

1. is kind, sympathetic and easy to talk to.

2. has the ability to explain clearly and in more than one way.

3. is completely fair and has no favorites.

4. is an enthusiastic and able mathematician.

5. has a good sense of humor.

6. treats pupils as responsible people.

7. uses methods which make mathematics more "alive" and interesting.

8. has patience with those who need more help.

9. understands pupils' problems with their work.

10. arranges work to suit the pace of the pupils.

be taught as an influence on teacher, pupil and observer perspectives this was not possible, simply because the structure and composition of the subject matter was rarely, if ever, explored in any depth by teachers or pupils. There was no way, therefore, to consider how pupil learning with respect to any mathematics actually proceeded and how it was influenced by cognitive as well as noncognitive factors. Mathematics educators now recognize that pupils make responses which are rarely random and illogical (although perhaps "wrong" from the perspective of the teacher). Pupil answers are thus usually based on quite rational and consistent abstractions from their learning experiences. These "alternative mathematical frameworks" need to be considered along with the more general motivational influences on pupil response. We need, therefore, in our research to look at how differences in both pupil conceptions and expectations might influence the effectiveness of mathematics teaching.

The Teacher Perspective

Research in the U.S. indicates that teachers "can and do make a difference and the differences are large enough that teachers of varying impact can be identified and studied" (Good, Grouws, & Ebmeier, 1983). Evidence suggests that pupils may fail to distinguish crucial mathematical issues introduced by the teacher from more peripheral aspects or fail to detect

Table 2

Ranking of Desirable Characterisitcs

in a Mathematics Teacher

2. has the ability to explain clearly and in more than one way

8. has patience with those who need more help

9. understands pupils' problems with their work

*

10. arranges work to suit the pace of the pupils

7. uses methods which make mathematics more "alive" and interesting

1. is kind, sympathetic and easy to talk to

*

6. treats pupils as responsible people

5. has good sense of humor

4. is an enthusiastic and able mathematician

*

3. is completely fair and has no favorites

*The gap between the clusters of factors signifies quite a substantial "jump" in the rating; the cluster of three factors 2, 8, 9, for example, was seen as being of much greater importance than cluster 10, 7, 1.

coherence in the mathematics presented. These confusions can sometimes be traced to weakness in the instructional process. Good and Grouws (1986) have suggested criteria to assist teachers in their planning of the development of student learning in relation to a particular mathematical topic or topics. They recommend, for example, that attention be given to prerequisite knowledge, relationships, and variety of representation, and that consideration be taken of the potential generality of the concepts introduced within a particular context of application. We must also recognise, however, that a teacher's approach to his or her instructional role will be effected by a constellation of factors concerning conceptions of mathematical knowledge, views about mathematics teaching, and views of individual differences and differences between groups (with regard to class, gender, and ethnicity, for example). The self-fulfilling prophecy effect of teacher expectations has

Table 3

Frequency Table (N = 28) Summarizing

Pupils' Ratings of Ms. X

Factor	Very good	Good	Average	Bad	Very bad
1)	20	7	1	0	0
2)	24	4	0	0	0
3)	19	3	6	0	0
4)	23	5	0	0	0
5)	14	12	2	0	0
6)	18	7	3	0	0
7)	5	18	3	2	0
8)	23	5	0	0	0
9)	20	5	3	0	0
10)	7	14	7	0	0

(Hoyles, et al, 1985, p. 48)

been widely documented, and, if an individual pupil encounters expectations from a teacher that are consistent over time, it is possible that the teachers' perceptions and the pupil's own self-image will become more and more alike. It would appear from naturalistic studies using teachers' real expectations of their pupils that high and low teacher expectations are related to differential teacher behavior. Some researchers report that teachers communicate more frequently and in longer exchanges with pupils deemed to be bright than with pupils deemed to be less able (e.g., Kester, 1969; Willis, 1969; Bryan, 1974; Lawlor & Lawlor, 1973). In addition, differences in the quality of interaction have been reported (Keddie 1971); for example, communications with allegedly bright pupils have been described as friendly, encouraging, and accepting (Kester, 1969), or as demanding and reinforcing of quality performance (Brophy & Good, 1970; Lawlor & Lawlor, 1973). These are compared to interactions with allegedly less bright pupils which have been found to be often negative (Willis, 1969), critical

(Cooper, 1977), control-oriented (Hargreaves, 1967; Rist, 1970; Cooper, 1977), or helpful and supportive without being challenging (Lawlor & Lawlor, 1973). It has also been noted that teachers often ignore responses from pupils deemed to be dull (Willis, 1969; Hargreaves, 1967). Cooper (1977) went further when he noted that teachers tended to perceive interactions with "low-expectation" pupils as time-consuming and that they often used criticism in order to control the timing and content of interactions with these pupils. Quality and quantity of interaction are also related to issues of race, class, and gender. As far as gender is concerned, it has been found that boys tend to monopolise teachers' time; they ask more questions and are asked more questions than girls (Mahoney, 1985; Spender, 1982; Stanworth, 1983), and they receive more praise and more blame than girls (Fennema 1980). More subtle gender-related differences in quality of interaction in mathematics classrooms are referred to later in this paper. All these studies suggest that teacher behaviour does tend to reflect teachers' expectations of their pupils, and these expectations can affect the personal qualities and source of motivation a teacher attributes to a pupil (e.g., Johnson, Feignebaum, & Weiby, cited in Waetjen, 1970).

In the Mathematics Teaching Project, we attempted to probe the teachers' perceptions of their pupils using repertory grids of the personal constructs that the teacher held of the pupil population in each project class. A triadic elicitation technique, by means of an interactive computer programme PEGASUS (Shaw, 1980), was used, the elements of the grid being pupils chosen by the teacher as "representative" of the project classes. The computer programme provided ongoing feedback and analysis and also constructed a cluster analysis of the constructs elicited. An important aspect of the elicitation in this research was that it was accompanied by in-depth interview in order to reflect on the grid as a whole and to discuss the meaning of pole names and an interpretation of the clusters. An example of the data is given below (first published in Hoyles, et al, 1985).

The details of our analysis of this repertory grid are given in Hoyles, et al., (1985). It is summarized as follows:

> Ms X's overall construct system separates into two general clusters that represent important distinctions made in her view of pupils. One cluster is based on perceived attitude and one on ability, particularly as this is manifested in interactions with her in the classroom. Both distinctions are very related to mathematics and the mathematics classroom as opposed to discriminations influenced by more general pupil characteristics. Thus for Ms. X, the pupil's "relationship" with the subject, mathematics, and their relationship with her as a mathematics teacher are of great significance. (p. 48).

Our examination of all grids of all the teachers in the Project reveal variations in both subject specificity and in the importance accorded to the development of personal qualities as opposed to mathematics expertise. This former emphasis was to a certain extent surprising. It would seem to

Table 4

Ms X's Personal Constructs of Her Pupils

Order in which constructs elicited	Bi-polar constructs	Comments
1	Consistent effort -- Avoids effort	Concerned with reaction to "challenge" in mathematics, i.e., determined effort or giving up.
6	Interested -- Lacks interest	Concerned with interest in mathematics and mathematical activity.
7	Co-operative -- Unco-operative	Concerned with the extent of co-operation with the teacher as a "manager" of the class.
5	Quiet -- Exuberant	Concerned with a perceived per- sonality trait of the children.
3	Accepting -- Questioning	Concerned with how the pupils "received" the mathematics introduced in the class.
2	Teacher dependent -- Flair	Concerned with autonomy and risk-taking in mathematics.
4	Anxious -- Confident	Concerned with pupil self-image of mathematical ability.
8	Neat -- Untidy	Concerned with the presentation of mathematics in written form.

Table 5

CONSTRUCT CLUSTER

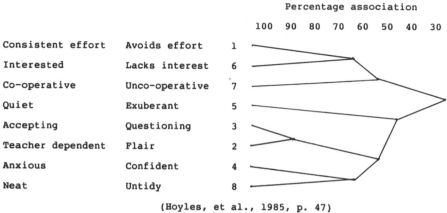

			Percentage association
			100 90 80 70 60 50 40 30
Consistent effort	Avoids effort	1	
Interested	Lacks interest	6	
Co-operative	Unco-operative	7	
Quiet	Exuberant	5	
Accepting	Questioning	3	
Teacher dependent	Flair	2	
Anxious	Confident	4	
Neat	Untidy	8	

(Hoyles, et al., 1985, p. 47)

reinforce the importance of taking into account affective factors when considering mathematics teaching.

The elicitation procedure described above produced rich descriptive data, but more work needs to be done in refining the technique so that it can be used in larger experimental studies. Such modifications have recently been proposed by Postlethwaite and Jaspars (1986) in their research with physics teachers where a critical triad procedure was adopted designed to elicit constructs which behaved as favorite or as disqualifying cues in a teacher's assessment of a pupil's potential. Attribution theory may also provide insight into the way teachers explain the achievement results of their pupils; that is, whether the cause is perceived to reside within or outside the individual concerned (its locus); whether it is perceived as temporary or as relatively enduring (its stability); and whether it is subject to volitional influence (its controllability) (Weiner 1980, 1982). This work would need to be undertaken with reference to the gender of teacher and pupil.

Another influence on teachers' approaches to their role is their view of what is meant by "knowing mathematics" or, even more fundamentally, what is meant by "mathematics." Bishop and Nickson (1983), after their review of research on the social context of the mathematics classroom, pointed out: "There is much public concern about teachers' knowledge of mathematics, but research into the social context makes plain the need to focus more on attitudes and perceptions of teachers *with respect to the mathematical content of the curriculum*" (p. 62; emphasis added). This sort of plea has also been made by Thom (1973), who wrote ". . . all mathematical pedagogy, even if scarcely coherent, rests on a philosophy of mathematics" (p. 204), and Hersh (1979), who pointed out that "the issue then, is not, what is the best way to teach, but what is mathematics really all about . . . controversies about high school teaching cannot be resolved without first confronting problems about the nature of mathematics" (p. 18). Confrey (1980) has also described how each theory of knowledge brings with it a commitment to a particular set of curriculum decisions.

Some studies now exist which try to explore those ideas; for example, Thompson (1984) examined teachers' beliefs about mathematics and mathematics teaching and the connections with teachers' behavior using case studies of three junior high school teachers. She concluded that ". . . teachers' beliefs, views, and preferences about mathematics and its teaching, regardless of whether they are consciously or unconsciously held, play a significant, albeit subtle, role in shaping teachers' characteristic patterns of instructional behavior" (pp. 124-125). Further work in this area has been undertaken by Brown and Cooney (1985) and Jones, Henderson and Cooney (1986).

Lerman (1986) has argued that an analysis of perspectives of the nature of mathematics must be undertaken from a theoretical point of view before empirical work is undertaken. Such an examination led him to propose the following two separate streams: "Firstly, a tendency towards seeing mathematics as based on indubitable, value free universal foundations which may not yet have been completely determined; and secondly, a view of mathematics as a social invention, its truths being relative to time and place" (p. 2). Lerman further suggested that:

> One can distinguish between two ways of teaching, which reflect this separation, the first being a "closed" view, whereby the teacher is the possessor of knowledge which is to be conveyed to the recipients, the pupils. The second is concerned with enabling pupils to be actively involved in the processes of doing mathematics, encouraged by "open" teaching, in the sense of the teacher working from the ideas and concepts of the pupils. (p. 3)

It is arguable, however, whether an individual's beliefs about the nature of mathematics are readily accessible to an outsider or even to the individual concerned who may only have a very patchy insight into his or her own belief system. Lerman, in fact, found that despite variation in teachers' views of mathematics, together with differences in background experience and training, there was considerable uniformity in ways of teaching mathematics. This seems to indicate that factors other than teachers' views of mathematics more crucially affect teaching behavior; in particular school context factors, for example, general ethos of the school, texts used, and the "philosophy" within a mathematics department. Thus, there seems to be an inevitable translation of a variety of epistemological perspectives into a "common" set of teaching objectives. This, in my view, leads us to an interesting new line of inquiry; that is, that school mathematics (as opposed to mathematics) cannot be examined independently of educational discourse.

TEACHING SCHOOL MATHEMATICS

In this section I wish to draw attention to the importance of the social influences on teaching and learning in mathematics classrooms and to how these need to be taken into account in any interpretation of classroom events. Learning does not take place in a social vacuum and is affected by institutional, cultural, and historical forces. When, for example, we observe that there is little negotiation of mathematical meaning between teacher and pupils, we must question whether this is due to lack of initiative or confidence on the part of the pupils, to lack of diagnostic skill on the part of the teacher, or to constraints built into the classroom situation. Thus we must consider the possibility that perceived behavior may not be the result of individual decision making but may be a consequence of social context. Mathematics has particularly powerful social connotations, especially in school, where it is generally accorded high status. One might therefore be led to enquire what school mathematics actually is. A good indication of what constitutes school mathematics can be obtained by taking a look at school textbooks and syllabuses, which exhibit a very high degree of standardization across different countries. In the terminology of Bernstein (1971) mathematics can be considered as having strong classification and strong frame (Note 1). This has consequences as to who is regarded as "having knowledge" and how this knowledge is acquired. School mathematics tends to be defined "from the outside" and dominated by the examination system. Because of the socially defined claims on school mathematics, it can even be regarded as a subject distinct in itself; that is, distinct even from mathematics as practised in a scientific community. School mathematics has its own ways of working (for example, step by step and hierarchical) and even its own concept names and symbols (cf. Dorfler & McLone, 1986). Thus what is termed by the French mathematical educators as a "didactical transposition" has occurred in which an "adaptive treatment of mathematical knowledge" takes place in order "to transform it into knowledge to be taught" (Balacheff, 1984, p. 35). School mathematics can therefore be conceived as a disciplinary matrix (Note 2) recontextualised from academic mathematics through its articulation into an educational discourse.

Social interaction between teacher and pupil both in class teaching and in small group/individualized work in mathematics still tends to reflect and reinforce a transmission model of teaching and learning where knowledge and expertise is assumed to reside with the teacher. Bauersfeld (cited in Lorenz, 1980) has identified a consistent pattern of "funnel-like" communication in mathematics classrooms where teachers' questions increasingly limit the scope of possible pupil response and contribute to pragmatical practical behavior on the part of the pupil; for example, to guess what is in the mind of the teacher (Voigt, 1985, p. 99). Teachers and pupils thus seem to "connive together" both in insisting on a repetitive and highly structured curriculum and in avoiding the exposure of any differences in understanding or interpretation of a problem. As Lorenz (1980) noted:

> There is no other subject in which the teacher is so tempted to misinterpret a (numerically) correct student response as an insight into the underlying problem structure; and nowhere is the student more willing to accept overt or covert prompts in order to conceal problems in understanding." (p. 18)

The following is an example I observed recently. A pupil had been asked to substitute $x = 4$ into the expression $3x - 8$. The pupil was stuck and asked for help:

Teacher: What is 3 times 4?

Pupil: 12

Teacher: What is $12 - 8$?

Pupil: 4

Teacher: OK then?

Pupil: I just write 4 then? OK fine!

(Hoyles, 1985, p. 26)

In the pattern of interaction illustrated here the teacher seems to be compelled to ask questions which the pupils can answer and which can then be taken as evidence that the knowledge has been "acquired." But, as Brousseau commented, "the meaning of this knowledge is entirely dependent on what is left as a activity to be done by the pupils" (Brousseau, 1984, p. 11) (here not very much!). The meaning will also depend on whether the interaction bears any relationship to the pupil perspective on the problem. Here, for example, imagine the impact of the teacher's questions on a pupil who had the commonly held view that $3x$ was "thirty-something!" However, Brousseau (1984) has suggested that when teachers' questions are used in this way, that is, to transform, deform, or even evaporate the knowledge, "this transformation is not the teacher's fault, rather it is dictated by the compulsions he [sic] cannot escape" (p. 111). Brousseau argues that under the constraints of what is termed the *didactic contract*, a set of rules which organize the relations between the content taught, the pupils, and the teacher, "the teacher is led in certain circumstances to empty the learning situation of all cognitive content" (p. 112). Thus the pupils obtain solutions by interpreting the didactic contract and not by involving themselves in the task. The point to be made here is that this type of interaction can be interpreted as an outcome of the relationship between the teacher and pupil in the present situation in mathematics classrooms *despite individual teacher's intentions and preferences. If this is the case it would seem important to consider these social constraints when undertaking research into mathematics teaching.*

The classroom situation is not, of course, completely determined, and some pupils—mainly boys—manage to become insulated from what might be termed the dogmatism of school mathematics and the recursive cycle of dependency engendered by teacher-pupil interaction (Grieb & Easley, 1984). Our first interpretation of interaction patterns in mathematics classrooms observed in the U.K. suggests subtle yet important differences in quality and quantity of interactions between the teacher and individual pupils or groups of pupils. These can be seen to relate to the gender of the pupil, although that of the teacher has also been noted as a critical factor. Girls seem more likely to be encouraged towards a rather careful step-by-step unadventurous approach to mathematics by teachers who do not expect them to be creative in their mathematical work. It has also been observed that teachers (particularly at the elementary level) have an overwhelming tendency to describe girls as "mature" and boys as "having potential," or, put another way, girls as "not *really* bright" and "working to their full potential" and boys (with similar scores on mathematics tests) as "*really* bright" but "underachieving!" (Walden & Walkerdine, 1985). (For further discussion of this research area see Leder, 1985 and Fennema, 1985).

Similar observations have been made elsewhere. In an overview of all the research in the last decade on the interactions of male and female teachers in elementary schools, Brophy noted that "studies that have examined subtle qualitative aspects of teacher behavior suggest that teachers may be socializing boys relatively more toward self reliance and independent achievement striving while socializing girls relatively toward conformity and responsibility." (Brophy, 1985, p. 30). As far as the secondary school is concerned, Brophy found that the picture was confused, but if a focus is taken on *mathematics* classrooms (as opposed to all classrooms at secondary level) there could be said to be some support for the findings of Becker (1981), who reported that the environment of a geometry class was, for boys, "supportive academically and emotionally" but for girls "one of benign neglect" (Becker, 1981, reported in Brophy, 1985, p. 312).

THE PROBLEM

The Cockcroft Report (Committee of Inquiry into the Teaching of Mathematics in Schools, 1982) in the U.K. suggested that it is not possible to indicate a definitive style for the teaching of mathematics; and approaches to the teaching of a particular piece of mathematics need to be related to the topic itself and the abilities and experience of both teachers and children. It further pointed out that methods which may be extremely successful with one teacher and one group of children will not necessarily be suitable for use by another teacher or with another group of children. Nevertheless it was proposed that certain elements should be present in successful mathematics teaching to pupils of all ages.

Mathematics teaching at all levels should include opportunities for:
• exposition by the teacher;

- discussion between teacher and pupils and between pupils themselves;
- appropriate practical work;
- consolidation and practice of fundamental skills and routines;
- problem solving, including the application of mathematics to everyday situations;
- investigational work.

Although there are some classrooms in which the teaching includes, as a matter of course, all the elements which are listed above, there are still many in which the mathematics teaching does not include even a majority of these elements. The implications of this fact are considerable. It is essential that standards of mathematics teaching should be improved overall. The extent to which this is accomplished will depend in large degree on the extent to which those who teach mathematics are able—and can be helped—to work in the way advocated above. (Committee of Inquiry into the Teaching of Mathematics in Schools, 1982, p.71)

Exhortations of this nature are now affecting, at least superficially, what is going on in classrooms in the U.K.; but will mathematics teaching be more effective? Encouraging pupils to talk about their methods and perceptions, to justify their strategies of exploration and proof, and to learn from each other implies a major shift in the social relations in the classroom which may not be acceptable to some teachers, either because they are concerned about management issues (given class sizes and general teacher stress) or because they do not accept the educational rationale for the change. The recommendations fail to recognize the actual nature of mathematics classrooms where "doing mathematics" tends to be subordinated to the demands of schooling as described earlier. One must ask whether it is possible to produce change by extending the range of teaching activities in the classroom or whether a much more fundamental shift in perspective is required which involves altering the context of teaching and learning. We know that performance on a problem can be quite dramatically affected by the context in which the problem is embedded (APU 1985), the way it is posed, and the method of assessment used (Note 3). We are also becoming increasingly aware of the different modes of doing mathematics which take place in everyday work settings and how these have very different achievement results (cf. Rogoff & Lave, 1984); Carraher, Carraher, & Schliemann, 1985). There is, thus, a growing body of research which suggests that mathematical understandings and strategies tend to be associated with *specific* experiences, one of which is schooling. There is a gap between intuitive mathematics displayed in everyday settings and facility exhibited in the school situation where, in particular, formalization is required. I would therefore like to suggest that effective mathematics teaching should attempt to capitalise on the mathematics children use for instrumental ends. Pupils can and do apply mathematical ideas and operations as tools in situations that are meaningful to them; that is, outside the disciplinary matrix of school mathematics and within their own everyday practices. These activities are generally routinised and mechanical; they are not necessarily describing the world in terms of mathematical models or symbolic operations. Effective mathematics teaching might, however, seek ways to provoke awareness in an explicit sense of the embedded mathematical concepts and relationships within such functional activity.

A MODEL FOR LEARNING AND TEACHING MATHEMATICS

With this background in mind, a model for learning mathematics has been proposed which involves the dynamically related components of *Using, Discriminating, Generalizing and Synthesizing* (UDGS) (also described in Hoyles, 1986). I would like to take this model as a starting point when thinking about teaching mathematics. The essence of the model is: First use mathematics in situations in which its power is appreciated (although there might, at this stage, be little more than a syntactical or procedural level of understanding of what is being used), then rewrite in explicit form what was previously implicit and break down and reflect upon procedures that previously only operated in totalities, and, finally, attempt to apply these structures in new domains and to make links with other learned procedures.

The components of the model are as follows:

Using, where a concept is used as a tool for functional purposes to achieve particular goals.

Discriminating, where the different parts of the structure of a concept used as a tool are progressively made explicit.

Generalising, where the range of application of the concept used as a tool is consciously extended from a particular to a more general case.

Synthesising, where the range of application of the concept used as a tool is consciously integrated with other contexts of application; that is, where multiple representations of the same knowledge in different symbolic codes derived from different domains are reformulated into an integral whole.

Movement between all the parts of the UDGS model is achieved through the strategies of conjecturing, testing, and debugging. Experimentation and willingness to try out ideas is at the heart of the UDGS model for learning mathematics, and the teacher's role is to encourage this mode of working and to provoke reflection on what has taken place. In this way, teachers assist their pupils in restructuring knowledge from its initial basis within "theorems in action" (Vergnaud, 1982) to more abstract cognitive structures; that is, they make the transitions through the components of the UDGS model.

The semantic level at which pupils will engage in any mathematical activity is subject to social and affective influences as well as to their level of cognitive development. As far as the latter is concerned, a major thesis of Vygotsky (1978) is of relevance particularly as we are concerned here with the role of the teacher. Vygotsky proposed that every specific state of a pupil's development is characterized by an actual developmental level and a level of potential development. The pupil is not able to exploit the possibilities at the latter level on her own, but can do so with educational support. Thus teaching should provide "scaffolding" (Wood, Bruner, & Ross, 1979) for voyaging into the next level of intended learning. These ideas lead us on to think about a model of teaching based on the UDGS model for learning which *does not necessarily lead to conflict between the learner's autonomy and pedagogic guidance.*

A Way Forward

Given the previous analysis in this paper, it would seem important that studies of mathematics teaching take into account teacher and pupil perspectives, the mathematical content, and the nature of the classroom environment. So how do we cope with such complexity in research? First, we need to develop further our theories so that we can more precisely identify and interpret what is observed. Second, in trying to explore the relationship between our theories and classroom practice, it would seem desirable in order to cope with the complexity of our task to focus on rather specific situations; that is, design, try out, and evaluate situations which optimize the possibility of some mathematical knowledge being constructed by pupils, taking into account the different starting points of the pupils as well as their different ways of working, levels of confidence, and task acceptance. So let us start by trying to find ways of embedding mathematics within rich activities in which pupils can be engaged. This idea relates to Papert's conception of a microworld of reality: "a constructed reality whose structure matches that of a given cognitive mechanism so as to provide an environment where the latter can operate effectively" (Papert, 1984). It also relates to the need identified by Romberg and Carpenter (1984) for "dynamic models that capture the way meaning is constructed in classroom settings on specific mathematics tasks" (p. 868). Our conception of a microworld, spelt out in detail elsewhere (Hoyles & Noss, 1987), incorporates components relating to pedagogical guidance as well as to pupil conceptions and mathematical knowledge. This represents a first step in trying to link teaching strategies more directly with pupil activity within specific learning environments. I would also like to suggest that the role of the teacher in such learning environments is:

- to specify the characteristics of the task to be performed;
- to structure the investigation and focus attention on points of the activity which are critical in making explicit for the pupils the nature of the "tools" they are using in the activity;
- to orientate the pupils to the development of their own metacognitive processes, that is, to encourage prediction, reflection, and evaluation;

- to facilitate the integration of possibly contradictory perceptions within a pupil's mind;
- to move a pupil from the performance of operations in context- specific "special cases" towards generalization and some form of symbolic representation.

I must now, though, return to what for me is still a fundamental problem: How can effective mathematics teaching as I have described it take place given what we know about the constraints of the classroom? We know that teachers and pupils tend not to search together in a genuine and open way to uncover mathematical meaning. We know, for example, that pupils want teachers to "make it easy" or "tell them the way," and we have to recognize the powerful influences on teacher practice which almost compel an algorithmic approach. We need to find a significantly different mode of education and practice in our classrooms, new roles for teachers which they value and which they see as significant for the mathematics learning of their pupils. What is required is a powerful catalyst for change in order to "break the mould." The advent of the computer may be just such a catalyst, although I am loathe to see it as a panacea. However, discussion of the role of computers in mathematics education has revitalized educational debate and the time is ripe for many "taken for granted" views and assumptions to be looked at afresh.

It is not appropriate for me to go in any depth into the potential of the computer for learning mathematics. For the purposes of this paper, however, I do wish to claim that it can engage pupils in mathematical activity, it can provide informative feedback during which "bugs" can be treated constructively, and with a appropriate selection of computational tools, it can focus the learner's attention on specific mathematical ideas or processes. Thus, the tools can "provide the learner with 'a representational system tailor-made for understanding a particular mathematical structure" (Hoyles & Noss, 1987). Of more importance here is the potential effect of computer use on the culture of the mathematics classroom; that is, the possibility of making the mathematical ideas and processes used in school more "functional" for the pupils, more in their control, and more accessible to them and to their teachers. As Weir (1987) has claimed (in the context of Logo programming), "the most exciting part of the educational computing enterprise will be its effect on classroom culture: on attitudes and atmosphere, on the patterns of intervention, and on the location of control in the classroom." (p. 246).

We are convinced in our present research (Hoyles, Noss, & Sutherland, 1986) (Note 4) that many of the important aspects of pupil-teacher relationships, teaching styles, classroom organization, and pupil learning can be highlighted by a consideration of appropriate computer use. We have accumulated examples of teachers whom we have watched re-examine their attitudes to teaching and pupils and re-define their teaching styles as a result of using the computer in their classroom. Not being the sole authority gives teachers the freedom to become living role models to their pupils on "how to learn;" thus pupils are able to learn by seeing this process in action. Teachers also can find more "space" to observe their pupils. As Carmichael, Burnett, Higginson, Moore, and Pollard (1985) concluded after their research with Logo:

> All the teachers in our study had been surprised at how much children of all abilities and age levels were able to learn—they were able to discover this because they had the opportunity to look at children interacting with computers, . . . This significantly altered their way of "seeing" children as learners: old assessments and categorizations of children broke down and old teaching strategies were being reassessed. They discovered that children did not learn only by incremental steps from simple to more complex ideas but often worked "backwards" from the more complex. (p. 366)

Thus we are now beginning to see how computers can provoke change in what a teacher sees as her role, how she can be freed to stand back and act as a facilitator and forego the immediate gratification of "telling the answer."

There are, of course, enormous variations amongst teachers, and the complexity of this issue must not be underestimated; in particular, it is important to recognise that the computer does not have influence by and of itself. The effect of its presence depends on the software used and how it is incorporated into the classroom milieu. It is still open to question as to whether the computer will facilitate a wide range of teaching and learning styles or whether "the computer will simply be one more potential tool that is ignored or misused by our schools" (OCDE, 1986, p. 3). Many teachers are very uncertain about the role of the computer in their teaching and are, quite rightly, wary of accepting this latest "fashion." New approaches can impose extra workload, can be threatening and can engender feelings of guilt or inadequacy. The unthinking use of the computer or its use in situations where other materials might be more appropriate is already leading to rejection of the new technology. However, it *is* the case that the computer can be used to create a new type of environment for learning mathematics, an environment which is activity based in the way described earlier and which supports pupil investigation. Put another way: The computer with appropriate and powerful software has the potential of changing the didactical relationships in schools.

Mathematics teaching, in my view, is about facilitating the learning process of pupils. Good teaching requires a combination of subject competency, a flexibility of teaching style and strategy, and a concern for the emotional and social as well as the cognitive needs of pupils. This requires the use of a range of teaching styles and a focus on pupil conceptions and ways of working as well as on mathematical content. I would like to suggest the we might apply the UDSG model of learning mathematics to "learning teaching mathematics;" for example, first do it in a functional way to achieve a planned outcome in terms of pupil learning, then reflect upon and analyse what has happened and attempt to formulate more general strategies. In particular it will be important to try to interpret the surprises: If pupils work in a different way from that expected by an a priori task analysis, why is this the case? How can these interpretations be incorporated into a more complex task analysis for use on subsequent occasions? This sort of work will clarify methods of teaching specific knowledge and, perhaps, provide a basis for abstraction of more generalized competencies.

An important aspect in an approach as described above is that teachers become aware of and regulate their own teaching strategies; that is, become researchers of their own practice. For this to occur teachers need to develop their own self confidence with respect to mathematics and mathematics teaching. This poses considerable problems both for research on mathematics teaching and for teacher education. The way forward on both these fronts, in my view, is *to work with teachers as research collaborators*. We are at present doing this in London; that is, working with teachers in the construction of computer-based mathematical microworlds and their evaluation in schools. This evaluation takes into account not only pupil learning but the teaching process by which this learning takes place. Thus the development and implementation of the microworlds allows the teaching of mathematics to become concretised, an object for reflection by the teachers themselves. It has provided a way to synthesize previous research on effective mathematics teaching with the craft knowledge of practising teachers, with research on how mathematics is learned, and, how school mathematics is socially enacted. In particular, we have asked whether computer-based work can provoke a change in classroom relations and how it might provide a means of synthesising a structured with an exploratory approach to mathematics teaching. We have faced together issues of classroom organisation and management, motivation and expectation, learning needs and styles, and the meaning of mathematical knowledge from a pedagogical point of view and from the point of view of pupils.

We, as researchers, still have many questions to investigate quite aside from teacher education. We need to develop further our tools of analysis of the teaching/learning situation, taking into account teacher, pupil, and knowledge as well as the interactions between these three poles in a classroom context. In formulating research questions in this area I feel we must resist the temptation to cope with their complexity by ignoring some of the variables, for example, by focusing solely on cognitive aspects, on attitude of the teacher, on classroom practices, and so forth. All such approaches are doomed to be oversimplified. They lead us to think that we are investigating something important only to realise later, when confronted by the complexities of practice, that our findings are not so significant after all! In our endeavor we need to develop didactical tools which construct a fruitful dialectic between pupils' everyday intuitive "mathematical" activity and their work in school mathematics. Whilst recognising the essentially paradoxical nature of our task, we must seek ways to promote pupil autonomy in a school situation where pupils cannot really be autonomous. We should aim to provide teachers not with recipes as to "how to teach" but with a theoretical framework through which they can construct and evaluate their own teaching. We should perhaps try to capitalise on the very different education systems in different countries and to pool our ideas and ways of working. Investigation in contrasting perspectives can help to highlight the theoretical issues involved in teaching mathematics. Similarly, we need mathematics specialists to work alongside education researchers and teachers. Only by such means can we hope to achieve some radical new thinking on an age-old problem.

NOTES

Note 1

"Classification refers to the degree of boundary maintenance between contents . . . Frame refers to the degree of control teacher and pupil possess over the selection, organization and pacing of the knowledge transmitted and received in the pedagogical relationship" (Bernstein, 1971, p.49-50).

Note 2

The term *disciplinary matrix* is taken from the work of Kuhn (1970) who describes it as that which members of a scientific community shape. It comprises at least four kinds of elements: symbolic generalizations, metaphysical paradigms, values, and exemplars.

Note 3

Styles of teaching in which pupils work cooperatively and not directly in competition and different modes of assessment, such as the use of multiple-choice or essay-type questions, may produce differing performance outcomes (cf. Murphy, 1980).

Note 4

The Microworlds Project (Hoyles, Noss, & Sutherland, 1986) is involved in the development, implementation, and evaluation of mathematical microworlds based largely upon the use of "content-free" computer applications (Logo, spreadsheets, database packages). The work is undertaken collaboratively with teachers as part of a programme of in-service education.

REFERENCES

Assessment of Performance Unit (1985). *Retrospective report: A review of monitoring in mathematics 1978-1982*. London: Her Majesty's Stationery Office.

Balacheff, N. (1984, November). French research activities in didactics of mathematics, some key words and related references. *Theory of mathematics education, Occasional paper 54*, (pp. 33-34), Institut fur Didactich der Mathematik (IDM), Universitat Bielefeld.

Barrow, R. (1986, November). Empirical Research into teaching: the conceptual factors. *Educational Research*, *28*(3), 220-230.

Becker, J. (1981). Differential teacher treatment of males and females in mathematics classes. *Journal for Research in Mathematics Education*, *12*, 40-53.

Becker, W. C. (1971). Teaching concepts and operations, or how to make kids smart. In W. C. Becker (Ed.), *An Empirical basis for Change in Education* (pp.401-423). Chicago: Science Research Associates.

Bernstein, B. (1971). On the classification and framing of educational knowledge. In M. Young (Ed.), *Knowledge and Control* (pp. 47-69). London: Collier Macmillan.

Bishop, A. J. & Nickson, M. (1983). *Research on the Social Context of Mathematics Education*. Windsor, UK: NFER-Nelson.

Bromme, R., & Juhl, K. (1984, November). *Students' understanding of tasks in the view of mathematics teachers, Occasional Paper 58*, Universitat Bielefeld.

Brophy, J. & Good, T. (1970). Teachers' communications of differential expectations for children's classroom performance: Some behavioral data. *Journal of Educational Psychology*, *61*, 356-374.

Brophy, J. & Good, T. (1986). Teacher behavior and student achievement. In M. C. Wittrock (Ed.), *Handbook of Research on Teaching* (3rd ed., pp. 328-375). New York: Macmillan.

Brophy, J. (1986, April). *Teaching and Learning Mathematics: Where Research should be going*. Paper presented at the annual meeting of the American Educational Research Association, San Francisco.

Brousseau, G. (1984, November). *The crucial role of the didactical contract in the analysis and construction of situations in teaching and learning mathematics*, *Theory of mathematics education*, *Occasional Paper 54* (pp.110-119). Institut fur Didaktich der Mathematik (IDM), Universitat Bielefeld.

Brown, C. A., & Cooney, T. J. (1985). *The importance of meanings and milieu in developing theories on teaching mathematics*. Paper presented at the Conference on Foundation and Methodology of the Discipline Mathematics Education, Bielefeld, West Germany.

Brown, J. S. (1984). *Process versus product—a perspective on tools for communal and informal electronic learning*. Unpublished manuscript, Xerox PARC.

Bruner, J. S. (1961). The Act of Discovery. *Harvard Educational Review*, 21-32.

Bruner, J. S. (1966). *Toward a theory of instruction*. Cambridge, MA: Harvard University Press.

Bryan, T. S. (1974). An observational analysis of classroom behaviors of children with learning disabilities. *Journal of Learning Disabilities*, *7*, 36-34.

Carmichael, H. W., Burnett, J. D., Higginson, W. C., Moore, B. G., & Pollard, P. J. (1985). *Computers, children and classrooms: A multisite evaluation of the creative use of microcomputers by elementary school children*. Canada: Queen's Printer for Ontario.

Carraher, T. N., Carraher, D. W., & Schliemann, A. D. (1985). Mathematics in the streets and in schools. *British Journal of Developmental Psychology*, *3*, 21-29.

Clark, M. C. & Peterson, P. L. (1986). Teachers' thought processes. In M. C. Wittock (Ed.), *Handbook of research on teaching* (3rd ed., pp. 255-296). New York: Macmillan.

Cobb, P. (in press). The tension between theories of learning and instruction in mathematics education. *Educational Psychologist*.

Committee of Inquiry into the Teaching of Mathematics in the Schools, (1982). *Mathematics Counts* (The Cockcroft Report). London: Her Majesty's Stationery Office.

Confrey, J. (1980). *Conceptual change analysis: Implications for mathematics and curriculum inquiry*. Institute for Research on Teaching, Science-Mathematics Teaching Center, Michigan State University, E. Lansing, MI.

Confrey, J. (1986, April). *A critique of teacher effectiveness in research in mathematics education*. Paper presented at the annual meeting of the American Educational Research Association, San Francisco.

Cooper, H. (1977). Controlling personal reward: Professional teachers' differential use of feedback and the effect of feedback on students' motivation to perform. *Journal of Educational Psychology*, *69*, 419-427.

Crosswhite, F. J. (1986, October). Better teaching, better mathematics: Are they enough?. *Arithmetic Teacher*, *34*, 54-59.

diSessa, A. A. (1982). Unlearning aristotelian physics: A study of knowledge-base learning. *Cognitive Science*, *6*, 37-75.

Dorfler, W. & McLone, R. (1986). Mathematics as a school subject. In B. Christiansen, A. G. Howson, & M. Otte (Eds.), *Perspectives on mathematics education* (pp. 49-98). Holland: D. Reidel.

Englemann, S. (1968). The effectiveness of direct verbal instructions on I. Q. performance and achievement in reading and arithmetic. In J. Hellmuth (Ed.), *The Disadvantaged Child* (Vol.3, pp. 339-361). New York: Brunner/Mazel.

Fennema, E. (1980). Teachers and sex bias in mathematics. *Mathematics Teacher*, *73*, 169-173.

Fennema, E. (Ed.). (1985). Explaining sex-related differences in mathematics: Theoretical models. *Educational Studies in Mathematics, 16*, 303-320.

Galton, M., Simon, B., & Croll, P. (1980). *Inside the primary classroom.* London: Routledge & Kegan Paul.

Good, T., & Grouws, D. A. (1986). *Developing in-service training programs for increasing understanding.* Columbia, MO: University of Missouri-Columbia.

Good, T., Grouws, D., & Ebmeier, H. (1983). *Active mathematics teaching.* New York: Longman.

Grieb, A., & Easley, J. (1984). A primary school impediment to mathematical equity: Case studies in rule-dependent socialization. *Advances in Motivation and Achievement, 2*, 317-62.

Hargreaves, D. H. ((1967). *Social relations in a secondary school.* London: Routledge & Kegan Paul.

Hersh, R. (1979). Some proposals for reviving the philosophy of mathematics. *Advances in mathematics, 31*(1), 31-50.Hoyles, C. (1982). The pupils' view of mathematics learning. *Educational Studies in Mathematics, 13*, 349-372.

Hoyles, C. (1985). *Culture and computers in the mathematics classroom* Bedford Way Paper. London: University of London Institute of Education.

Hoyles, C. (1986). Scaling a mountain—A study of the use, discrimination and generalization of some mathematical concepts in a Logo environment. *European Journal of Psychology in Education,* ———.

Hoyles, C., Armstrong, J., Scott-Hodgetts, R., & Taylor, L. (1984). Towards an understanding of mathematics teachers and the way they teach. *For the Learning of Mathematics, 4*(2), 25-32.

Hoyles, C., Armstrong, J., Scott-Hodgetts, R., & Taylor, L. (1985). Snapshots of a mathematics teacher—Some preliminary data from the mathematics teaching project, *For the Learning of Mathematics, 5*(2), 46-52.

Hoyles, C., Noss, R. & Sutherland, R. (1986). *An INSET programme developing computer-based microworlds for mathematics.* Proposal to ESRC, 1986.

Hoyles, C., & Noss, R. (in press). Synthesizing mathematical conceptions and their formalization through the construction of a Logo-based school mathematics curriculum. *International Journal of Mathematics Education in Science & Technology.*

Jones, D., Henderson, E., & Cooney, T. (1986). Mathematics teachers' beliefs about mathematics and about teaching mathematics. *Proceedings of the eighth annual meeting of the North American chapter of the International Group for the Psychology of Mathematics Education,* (274-279). East Lansing, MI.

Keddie, N. (1971). Classroom knowledge. In M. F. D. Young (Ed.), *Knowledge and control: New directions for the sociology of education* (pp.133-160). London: Collier-Macmillan.

Kester, S. W. (1969). The communication of teacher expectations and their effects on the achievement and attitudes of secondary school pupils (Doctoral Dissertation, University of Oklahoma, 1969). *Dissertation Abstracts International, 30*, 1434-1435.

Kuhn, T. S. (1970). *The structure of scientific revolutions* (2nd ed.). Chicago: The University of Chicago Press.

Lawler, R. W. (1981). The progressive construction of mind. *Cognitive Science, 5*, 1-30.

Lawler, R. W. (1982). Designing computer microworlds. *Byte, 7*, 138- 160.

Lawler, E. P. & Lawlor, F. X. (1973, January). Teacher expectations: A study of their genesis. *Science Education,* 9-14.

Leder, G. (1985). Sex-related differences in mathematics: An overview. *Educational Studies in Mathematics, 16*, 304-309.

Lepper, M. R. (1985, January). Microcomputers in education: Motivational and social issues. *American Psychologist,* 1-18.

Lerman, S. (1986). *Alternative views of the nature of mathematics and their possible inference on the teaching of mathematics.* Unpublished doctoral dissertation, University of London.

Lorenz, J. H. (1980). Teacher-student interactions in the mathematics classroom: A review. *For the Learning of Mathematics, 1*(2), 14-19.

Mahoney, P. (1985). *Schools for boys?* London: Hutchinson.

Murphy, R. J. L. (1980). Sex differences in GCE examination entry, statistics and success rates. *Educational Studies*, 6(2), 60-75.

OCDE (1986, October). *Information technologies and basic learning.* General report of international conference. Paris.

Papert, S. (1972). Teaching children to be mathematicians versus teaching about mathematics. *International Journal of Mathematical Education in Science and Technology*, 3, 249-262.

Papert, S. (1980). *Mindstorms: Children, computers and powerful ideas.* New York: Basic Books.

Papert, S. (1984). *Microworlds: Transforming education, mathematics in theory.* Occasional Paper, M.I.T., Boston, MA.

Postlethwaite, K., & Jaspars, J. (1986). The experimental use of personal constructs in educational research: The critical triad procedure. *British Journal of Educational Psychology*, 56, 241-254.

Rist, R. C. (1970). Student social class and teacher expectations: The self-fulfilling prophecy in ghetto education. *Harvard Educational Review*, 40, 411-451.

Rogoff, B., & Lave, J., (Eds.). (1984). *Everyday cognition: its development and social context.* Cambridge, MA: Harvard University Press.

Romberg, T. A. & Carpenter, T. P. (1986). Research on teaching and learning mathematics: Two disciplines of scientific inquiry. In M. C. Wittrock (Ed.), *Handbook of research on teaching* (3rd ed., pp. 850-873). New York: Macmillan.

Shaw, M. L. G. (1980). *On becoming a personal scientist.* New York: Academic Press.

Shulman, L. S., & Keislar, E. R. (1966). *Learning by discovery: A critical appraisal.* Chicago: Rand-McNally.

Skinner, B. F. (1968). *The Technology of Teaching.* New York: Appleton- Century-Crofts.

Spender, D. (1982). *Invisible women.* Writers and Readers Publishing Cooperative.

Stanworth, M. (1983). *Gender and Schooling.* London: Hutchinson.

Thom, R. (1973). Modern mathematics: Does it exist? In A. G. Howson (Ed.), *Developments in mathematical education. Proceedings of the Second International Congress on Mathematical Education* (pp. 194- 209). Cambridge: The University Press.

Thompson, A. G. (1984). The relationship of teachers' conceptions of mathematics and mathematics teaching to instructional practice. *Educational Studies in Mathematics*, 15, 105-127.

Vergnaud, G. (1982). Cognitive and developmental psychology and research in mathematical education: Some theoretical and methodological issues. *For the Learning of Mathematics*, 3(2), 31-41.

Voigt, J. (1985). Patterns and routines in classroom interactions. *Recherches en Didactique des Mathematiques*, 6(1), 69-118.

Vygotsky, L. (1978). *Mind in society.* Cambridge, MA: Harvard University Press.

Waetjen, W. (1970). The teacher and motivation. *Theory and Practice*, 9(1), 10-15.

Walden, R., & Walkerdine, V. (1985). *Girls and mathematics: From primary to secondary schooling* Bedford Way Paper 24. London: University of London Institute of Education.

Weiner, B. (1980). *Human motivation.* New York: Holt, Rinehart and Winston.

Weiner, B. (1982). The emotional consequences of causal attributions. In M. Clark & S. Fiske (Eds.), *Affect and cognition: 17th annual Carnegie Symposium on Cognition* (pp. 185-209). Hillsdale, NJ: Lawrence Erlbaum Associates.

Weir, S. (1987). *Cultivating minds.* Cambridge, MA: Harvard University Press.

Willis, B. J. (1969). The influence of teacher expectation on teachers' classroom interaction with selected children. *Dissertation Abstracts International*, 30, 5072A. (University Microfilms 70-07647).

Wood, D., Bruner, J. & Ross, G. (1979). The role of tutoring in problem solving. *Institute of Child Psychology and Psychiatry*, 17, 89-100.

Computer Usage In the Teaching of Mathematics: Issues That Need Answers

Janet Ward Schofield and David Verban

University of Pittsburgh

Anyone with even the slightest familiarity with the U.S. educational system in the 1980s is well aware of the incredibly rapid proliferation of microcomputers in schools at both the elementary and the secondary level. The magnitude of this change is, however, truly startling. For example, in the two-year period from the spring of 1983 to the spring of 1985 the number of computers in schools which had at least one in 1983 more than tripled. By the end of that period almost all secondary schools and five-sixths of all elementary schools in the U.S. had computers for use in instruction (Becker, 1986).

Although the remarkable rapidity with which microcomputers are being placed in schools is obvious, the effect of this change on teachers, on students, and on school systems is not. In fact, opinions vary dramatically on what impact computers can or will have. At one extreme are those who see computers as having the capability to revolutionize education in absolutely fundamental ways. For example, Walker (1984, p. 3) makes the rather startling claim that "the potential of computers for improving education is greater than that of any prior invention, including books and writing." Others take quite a different stance, emphasizing the inherent conservativism of the teaching profession with regard to pedagogical change and the failure of other highly touted educational innovations to brlng about far-reaching changes.

The fevered rush toward acquiring microcomputers means that many teachers now have available to them a tool with which they have little or no familiarity. Further, the fact that major expenditures are being made on computer hardware at a time when many school systems have been feeling a financial pinch has led many systems to skimp on teacher training and support services. In addition, it seems apparent that school systems sometimes buy computers in response to parental pressures or because they want to gain prestige by being at the forefront of a new trend (Taylor & Johnson, 1986) rather than because they have a vision of the educational goals the computers will help them achieve or of how the change process can be handled in a way that maximizes the potential benefits of using microcomputers in instruction while minimizing negative effects. Thus we would argue that more valuable at this point than extra hardware is reflection and research on precisely what schools can accomplish with microcomputers and how this can best be achieved.

The goal of this paper is to make a modest contribution to this general effort. Specifically, we will attempt to outline what we believe are some of the major issues with regard to the usage of computer technology in mathematics teaching that need further study. We have chosen to direct this paper to the teaching of mathematics for three major reasons. First, mathematics is the major subject for which computers are used in both elementary and middle schools (Becker, 1986). In high school, mathematics and business courses are the dominant domains, in addition to computer science, in which computers are heavily utilized. Many of the points we raise are applicable to teaching other content areas as well. However, just as recent research on learning has benefited from attention to the specific issues associated with learning particular subject areas, there are enough differences between teaching in different domains to warrant approaching the issue of computer usage in a domain-specific way. Second, it is clear that mathematics education in the U.S. is seriously in need of improvement. One need only look at recent cross-national comparisons to see how serious the situation is (McKnight et al., 1987). Finally, we have chosen to focus on mathematics teaching because many of the thoughts developed in this paper were stimulated by an intensive ethnographic study of computer usage in the teaching of mathematics, which will be described shortly.

USING ETHNOGRAPHIC RESEARCH TO DEVELOP A RESEARCH AGENDA

Of course one can raise the question of whether it makes sense to derive a research agenda from ethnographic research, which by its very nature is limited to one or a comparatively small number of sites. Those raising such an issue would be likely to point out the folly of assuming that the important issues in one site are necessarily those at another, the inherent subjectivity in many aspects of qualitative research, and the like. We answer this question in the affirmative, making no claims that this is the only, or even the best, single way to set a research agenda.

There are several very real strengths to an ethnographic approach as a means for stimulating thought about a research agenda. First, one of the defining characteristics of this approach is its flexibility and openness. Specifically, although such investigations start with some basic themes or questions in mind, the primary goal is to discover what is important in a given situation rather than to test a set of hypotheses which have been formulated in advance. Thus qualitative investigations can easily be shaped to explore unanticipated issues. Such an approach is highly desirable when little is known about the phenomenon under investigation, as is clearly the case with questions regarding the impact of microcomputers in educational environments.

Qualitative methods are also particularly well-suited to exploring the context in which the phenomena under investigation occur, to suggesting ideas about social processes, and to capturing with both vividness and subtlety the perceptions of the individuals being studied (Reichardt & Cook, 1979; Schofield & Anderson, 1987). Again, all of these attributes are particularly important for the topic to be discussed here.

Although relatively little is now known about the impact of microcomputers on various aspects of schooling, one thing that is clear is that the educational context is very important in determining if and how microcomputers will have an influence. For example, after a close look at computer usage in three rather different school systems, Sheingold, Kane, and Endreweit (1983, p. 431) concluded that "the effects of microcomputers on education will depend, to a large extent, on the social and educational context within which they are embedded." Similarly, it is clear that social processes in the school will play an important role in influencing how microcomputers influence teaching. For example, Becker (1984) found that the amount and type of microcomputer utilization in elementary school classrooms is clearly related to the relative importance of different actors (teachers, principals, or other administrators) in the acquisition of those machines. Finally, the emphasis that qualitative approaches put on questions of meaning is likely to be especially helpful in understanding the interrelation between teaching and microcomputer usage. For example, research by Sheingold, Hawkins, and Char (1984) demonstrates that teachers' interpretations of their subject matter and of available software are critical to understanding how microcomputers are utilized and thus what their impact is likely to be on both teaching and learning. Specifically, they discuss the results of a field test of a computerized simulation game, the manifest context of which concerns rescuing a whale trapped in a fishing net. Some teachers interpreted the software much as its designers had and utilized it to supplement their teaching of concepts such as degrees, angles, and vectors. Others saw the software as a game about boats and navigation and limited its use to free periods or after school hours.

The preceding discussion suggests that any research agenda relating to microcomputers in schools needs to keep in mind that microcomputer utilization can be conceptualized as a dependent as well as an independent variable. On the one hand, a whole variety of factors influence *whether* and *how* these resources will be utilized. On the other hand, utilization is likely to have *consequences* that we need to understand. In either case one caveat is vitally important: Microcomputer usage in and of itself is hardly a unitary, conceptually satisfying variable. Factors such as the purposes for which the computers are used (drill and practice, simulations, tutoring, etc.), the specific software chosen to achieve these ends, the ratio of students to microcomputers, and the physical location of the computers (classrooms vs. school libraries, etc.), all seem likely to influence how teachers and others respond to the new technology and how, in turn, the technology influences teachers and schools.

As will be discussed in more detail shortly, the ethnographic study upon which this paper is based focused on one particular usage of microcomputers—their utilization as intelligent tutors in geometry classes. There are a number of reasons why such a study is of particular interest. First, and most importantly, perhaps more than any other presently envisioned use of microcomputers, their use as intelligent tutors holds the promise of improving and transforming schooling as we know it today. Skilled human tutors can increase learning by as much as a factor of four (Anderson, Boyle, Farrell, & Reiser, 1984). Nonetheless, cost considerations keep schools from providing individual tutors to students. In fact, one teacher is typically responsible for twenty-five to thirty students at a time, resulting in very minimal amounts of individualized interaction between most students and their teachers. Furthermore, in most high schools teachers are responsible for teaching one hundred or more different students each day, making it hard, if not impossible, to develop the sort of intimate familiarity that a private tutor gains with a student's thought patterns and level of knowledge. Intelligent computer-based tutors have the ability to work interactively, to follow what a student is trying to do, to diagnose the difficulty a student is having, and to provide instruction relevant to that difficulty (Anderson, 1984). Although at the moment the cost of the development of the software as well as of the hardware required for this sort of activity is high, within five years it may well not be prohibitive for educational uses (Lesgold & Lesgold, 1984). Thus, the study of an intelligent computer-based tutor is the study of an innovation that is close to being a practical reality from a technical and a fiscal perspective as well as being a potentially revolutionary change from an educational perspective. As such it is a good foundation for raising questions about what we need to know in order to ensure that the potential of this development is realized and to understand how teaching and learning will be influenced by it.

The Research Site

Data gathering took place during a 2-year period (1985-1987) in a large high school located in an urban setting. The school, which will be called Whitmore High School, serves approximately 1300 students from very varied socioeconomic backgrounds. Approximately 55% of the students are black, 40% are white, and 5% are from other, primarily Asian, ethnic groups.

The school's faculty is about 80% white and roughly 55% male. The gender composition of different departments varies dramatically in ways that one might expect given traditional sex roles. For example, males comprise about 70% of the mathematics teachers as compared to about 30% of the English teachers.

During the first year of the study, the school had three main locations in which computers were concentrated. One of these was the room in which 10 Xerox Dandy Tiger computers were located. These highly sophisticated and very expensive computers were loaned to the school by the developers of an intelligent system for tutoring geometry proofswhich was being field-tested at the school. The second main concentration of computers was in the computer science lab which contained 11 Tandy 1000 microcomputers. The third concentration of computers was a group of 12 AppleIIEs used by students in the school's gifted program. In the study's second year, the business classes, which had formerly had only a small number of computers, received about 20 new microcomputers. There were also a small number of other classes with one or two computers.

Research Methods

The two major methods of data gatherlng were intensive and extensive classroom observation and repeated extended interviews with students and teachers. Administrators were also interviewed when appropriate. Classroom observers used the "full field note" method of data collection (Olson, 1976), which involves taking extensive handwritten notes during the events being observed. Shortly thereafter,these notes are dictated on to audiotape and then transcribed. Observers made the field notes as factual and as concretely descriptive as possible to avoid unwarranted inferences.

One clear problem with the use of such notes as a data base is what Smith and Geoffrey (1968) have termed the "two-realittes problem"—the fact that the notes as recorded cannot possibly include literally everything that has actually transpired. Hence, a source of potential bias is the possibility of selective recording of certain types of events. Although this problem is impossible to surmount completely in qualitative observation, there are some steps that can be taken to minimize its negative effect. For example, we found it useful to have two researchers observe a single setting. Discussion of differences between the two observers' notes helped to point out individual biases or preconceptions. Another technique useful in reducing the effect of such biases is to actively seek out data that undercut one's developing assessment of a situation. These techniques, plus a number of others discussed in recent books on qualitative research in educational settings (Bogdan & Biklen, 1982; Goetz & LeCompte, 1984), were employed to reduce the "two-realities problem." Fuller discussion of methodological details can be found in Schofield (1985).

During both years of the study, weekly observations were made in four geometry classes taught by Mr. Adams (a pseudonym, as are all names used in this paper). These classes, of varying ability levels, all utilized the computer tutor developed by Anderson and Boyle (1985) to work geometry proofs. Since the tutors were not introduced until January of the study's first year, we were able to do a complete semester of observations in Mr. Adams' classes before the computer-tutors were available. As an additional source of comparative data, we also made weekly observations in the classes of the two other teachers who taught geometry at Whitmore High but who did not utilize the tutor. One of these teachers occasionally used the computer lab set aside for gifted students with his class of gifted students, as did Mr. Adams before the arrlval of the more powerful machines in his room. In the study's second year, another teacher, Mr. Brice, used tbe computer-tutor with one additional class of students. Both this class and another geometry class of Mr. Brice's, which was taught using traditional methods, were observed weekly. The classes using the computer-tutor ranged markedly in size. Some were small enough so that each student had his or her own computer to work on. In others, students worked in pairs on the ten available machines.

We have observed roughly 250 hours of instruction in the classes mentioned above, almost 100 of which involved classes actually using the computer-tutors. The observation described so far allows comparisons (a) across different time periods within individual classrooms utilizing computer-based instruction, (b) between different ones of these classes, (c) between classes taught in the first year of the project and the second when the teacher was more familiar with the tutor, and (d) between classes utilizing the computer-based tutor and those which do not.

Although the study's core focus was on geometry classrooms, as described above, we also observed regularly in almost all other sites at Whitmore High School in which computers were used regularly for instruction. The purpose of observing in these other settings was to shed light on the question of which problems and issues are connected with microcomputer usage *per se* and which are specific to utilizing the computer as an intelligent geometry tutor. Thus weekly observations were made in three or more different computer science classes. Similarly, weekly observations were made in the "gifted" computer lab. Regular observations were also made in other classes, such as visual communications, office automation, and business practice, in which computers were available. A total of roughly 250 hours of observation were conducted in the settings just described.

Observers, no matter how omnipresent or insightful, are at a great disadvantage if they do not test their emerging ideas through direct inquiry with those whom they are observing. Because interviews can be so useful in providing the participant's perspectives on events, both formal and informal interviews were the second major data-gathering technique utilized in the research. All students from each of the computer-tutor geometry classes were interviewed during both years of the study.[1] Each year the first interview occurred before the students began using the tutors and focused on the students' expectations for the tutor, their perceptions of their teacher, and the like. The second interview, conducted after the students finished using the tutors, concerned students' reactions to the tutor and their observations on how using the tutors had changed their classes. The bulk of these interviews consisted of open-ended questions. A random sample of students from the traditional geometry classes were also interviewed twice in the study's first year. These students were asked questions about their classes and their teachers identical to the non-tutor questions asked of the computer-tutor students.

Both formal and informal interviews were conducted with Mr. Adams and Mr. Brice, the two teachers using the computer-tutors, throughout the course of the research. The formal teacher interviews were tape-recorded and transcribed, as were all student interviews. Additional interviews were conducted with virtually all other computer-using teachers, such as those who supervised the "gifted" computer lab, special education teachers, and computer science teachers.

Although observation and interviews were the primary data gathering techniques utilized, other techniques were employed when appropriate. For example, archival material, such as letters sent to parents about the computer tutor, internal school memoranda and announcements, and copies of the student newspaper, were carefully collected and analyzed.

ISSUES NEEDING FURTHER RESEARCH

Why and How Should Computers Be Used In Teaching Mathematics

The first and most obvious finding of our study was that, with the exception of the field-test of the computer-based geometry tutor, computers were rarely, if ever, used in teaching mathematics at Whitmore. Although board-level personnel spoke of having a computer-based remedial mathematics Computer Assisted Instruction (CAI) program in all high schools, no such program existed at Whitmore. Careful monitoring of the uses to which the various computer labs were put turned up no evidence of the utilization of computers in any math course other than geometry. Most geometry classes never used computers at all. The two exceptions to this were two "gifted" classes, one taught by Mr. Adams and the other taught by another white male, Mr. Trowbridge. Mr. Adams took his class to the gifted lab several times in the fall before the computer-based tutor arrived. Mr. Trowbridge took his students two or three times during the course of the year.

This lack of utilization of computers in instruction could not be attributed primarily to either lack of availability of machines or to teachers' unfamiliarity with them. The gifted lab with its 12 computers was, according to our best estimates, used about 15% of the time on an average school day. Although the computers in the computer science lab were more heavily utilized, the head of the math department, which was formally responsible for instruction in computer science, indicated that he would try to make the lab available to other teachers, especially the mathematics faculty, if they needed it. Since some of the math teachers also taught computer science, unfamiliarity with either the specific hardware available or with computers in general was hardly a satisfactory explanation for the lack of utilization of computers in mathematics instruction.

The contrast between the general failure to utilize computers in math instruction and their intensive use in the field testing of the computer-based geometry tutor raised what to us is a fundamental issue: Just *why* and *how* should computers be used in mathematics instruction? It makes little sense to bemoan the low level of utilization unless one has a clear idea of what the advantages of computer utilization might be. Anderson and his colleagues have outlined a clear rationale for utilizing individualized computer-tutors. Specifically, they point out that because the tutor constantly monitors and structures students' problem-solving attempts, works from a carefully developed set of ideal and "buggy" rules in guiding the student, and provides immediate feedback, it has the capability to provide important facilitators of learning that one teacher can not hope to provide for a typical class of students (Anderson, Boyle, & Yost, 1985). The field test at Whitmore was a two-pronged effort aimed at both improving the software and seeing if students using it learned more than those in control classes.

However, in general, math teachers at Whitmore appeared to have little conception of what parts of the curriculum might well be taught using computers and of when and how they should be used for drill and practice, for simulations, for their graphic capabilities, or the like. Since software development is primarily a for-profit enterprise, it seems likely that most software manufacturers will tend to produce products which can be easily marketed as performing a specific function. However, it seems unlikely that either they or practicing teachers will have the time or motivation to take a more global look at the whole issue and to produce what one might think of as a model of the use of computers in mathematics instruction that links the actual or potential characteristics and capabilities of different kinds of computers to the instructional methods and goals that teachers have in mathematics courses.

Of course, some efforts have already been made along these lines. Walker (1986) and others have given thought to how the various capabilities of the computer might be useful in teaching mathematics. Patrick Suppes has developed a computer-based curriculum spanning both the elementary and high school years. Other innovators like Seymour Papert, and the group at the University of Illinois that developed PLATO have also developed computer languages or curricula out of a vision of the particular role computers can play in education. However, an enormous amount of conceptual and empirical work remains to be done. This work involves addressing fundamental problems of values, with questions such as "What do we want our children to learn?" While such questions are old, the answers may be new as we strive to prepare students to function in a world that is rapidly changing and ever more heavily influenced by the burgeoning of technology. In fact, the consensus of those who have dealt with this issue recently appears to be that a major shift in the goals of mathematics teaching is needed— away, for example, from an emphasis on rote practice of calculation skills and towards an emphasis on problem-solving, estimation, and statistics, to name just a few areas that have previously received relatively little attention in the usual mathematics curriculum (Committee on Research in Mathematics, Science, and Technology Education, 1985; National Council of Teachers of Mathematics, 1980; Romberg, 1984; Willis, Thomas & Hoppe, 1985).

Having addressed this basic issue of what should be taught, the next set of fundamental problems concerns what part of this material can be taught better with the assistance of computers than without them. Here at least three sets of questions arise. The first deals, of course, with the issue of cost effectiveness. The second set of questions concerns those functions which computers may perform uniquely well. Here we need to enquire what special capabilities the computer has and whether these unique capabilities have an important role to play in mathematics education. Finally, attention needs to be paid to the side effects of computer usage in teaching mathematics on a broad range of organizational, social, economic, and intellectual outcomes. A striking though perhaps frivolous example of just how unintended and unexpected various side effects can be is the fact that Mr. Adams, the teacher most heavily involved in the field test of the computer tutor, complained that using the tutors gave him sore legs. Although he was in good physical condition, the squatting he did when consulting with individual students seated at their computers involved muscles that lecturing at the chalkboard clearly did not.

Barriers to the Utilization of Computer Technology In Mathematics Instruction

Let us assume that attention to the question of "Should computers be used in teaching mathematics?" yields at least a qualified "yes" as an answer. Let us further assume, perhaps less safely, that attention to the related question of how computers should be used also yields answers that do not require a major revolution of school organization as we know it today, although it may require significant changes in teaching. We would then argue that the next fundamental question that sets the course for a research agenda on the use of computers in teaching mathematics is "What are the barriers and facilitators of effective utilization of computers by teachers of mathematics?" The basic problem evident at Whitmore was that at this moment in time the disincentives to computer usage outweigh the incentives for most teachers. Our goal in this section of the paper is to outline some of these barriers and incentives to utilization. Two types of research seem needed. First, we would argue that more exploration of the nature and extent of each of these barriers and incentives is important. Second, but perhaps even more important, would be a wide range of applied research on how best to overcome these barriers and to utilize the incentives effectively.

Lack of Familiarity with Computers

Several of the mathematics faculty at Whitmore doubled as computer science teachers and hence were quite familiar with using such machines. However, for those teachers and administrators who were not familiar with computers, this situation posed a major barrier to utilization. For example, the coordinator for the gifted program, who was in a good position to encourage teachers and students to use the Apple IIEs in the gifted lab, which generally sat idle except at lunch when students used them primarily for educational games and word processing, expressed his frustration at not knowing much about computers and not being able to find good ways to change that situation. "Sometimes I sit down [at home] with a beer in one hand and a manual in the other, but it's pretty complicated to learn. . . . It just takes so much time to learn. . . . and I don't have it."

One issue that was frequently coupled with comments about lack of knowledge about computers was a sense that trying to use them exposed one to potentially embarrassing situations which undermined one's sense of competence. For example, Mr. Trowbridge, who indicated that he found it hard to "get the hang of computers," argued that because of their youth, students could pick up on computers more quickly. Thus, by attempting to use computers with his class he reversed the usual situation in which he was more in command of the knowledge needed to perform well in class than were students. Ms. Prentiss, who supervised the use of the gifted computer lab during lunch periods, clearly experienced a similar feeling of bewilderment and threat when first confronted by hardware and software problems which she had no training to handle.

> I've changed! You know that . . . I'm getting a lot better at it . . . My husband taught me phrases like "Hmmmm, looks like there's a bug in the program." I always assumed it was *my* fault at the beginning. And then . . . I realized it's not my fault. It's inherent in the system's hardware and software. . . . It's not because I'm a nincompoop. *Learning that . . . made a huge difference!* Everything else falls into place if I can hold on to that!

Mr. Miller, the coordinator of the gifted program, who was mentioned above, described this situation vividly in a conversation with a member of our research team, only this time the sense of lesser competence wasfelt in relation to a colleague.

> Mr. Miller next came back to his comments about Mr. East, a science teacher who is the school's foremost "hacker". He said, "He's a computer whiz. He's way over my head . . . A couple of times I've asked him to explain things to me, but it gets so complicated. He goes on and on and I just sit there and say "I gotcha . . . I got it. I understand," but I don't understand a thing!

The necessity for having someone on the spot who is familiar with both the hardware and software in use was obvious in both the geometry computer-tutor lab and in the gifted computer lab. In the former situation it took two or three adults to keep things running fairly smoothly, and even this level of staffing was sometimes not adequate to the task. This extraordinary level of staffing was clearly connected to the fact that the software utilized was still being developed and refined. But observation of the gifted lab suggested that a wide range of expertise was necessary to keep things running smoothly, even with fairly simple machines and widely used commercially developed software.

Ms. Prentiss, who bravely volunteered to supervise the lab as a "duty period" after having seen her husband accomplish a lot on their personal computer at home, had no formal training relating to computers except for a two-hour, school-system-sponsored workshop a few years previously. After the first semester she convinced Mr. East, one of the school's few real "hackers", to assist her for part of the lunch period. Even with two adults doing their very best, the 6 to 12 students who usually came could not count on having machines and software operating smoothly, as the following excerpt from our field notes indicates.

> The students . . . continue to have a lot of very nitty-gritty problems. Kathy can't get the printer going. . . . She's scowling and says in an annoyed tone of voice, "Please help me." Mr. East suggests several things, and after they try out four or five different approaches they finally get the printer working. Ms. Prentiss has been working with Sharon on word processing . . . For the last 10 minutes cries like, *"I don't believe it"* and *"Oh, no. Not again!"* have been emanating from both of them . . . Finally Ms. Prentiss calls Mr. East over . . . Sharon is clearly getting anxious, pacing around, picking her nails and the like. She takes her disk and inserts it in another computer hooked up to a different printer. She can't get this printer to work. . . . Ms. Prentiss rushes over to try to fix it saying, "I just don't believe it!" Ms. Prentiss comes over to me (the observer) and says "I feel like quitting this . . ." At this point Mark calls to Ms. Prentiss "I need help . . ." Ms. Prentiss puts her head down on the desk briefly. She looks at me with what appears to be a mixture of mock and real despair and trudges over to Mark. (Later in the same period) Dan is trying to use a printer which Mr. East thought he had fixed. Dan's essay comes out quadruple spaced. In addition, every single word is underlined. Ms. Prentiss looks at it and breaks into almost hysterical laughter. Dan looks annoyed. Ms. Prentiss says "I'm sorry, this is just too much— too, too much! . . ." Mr. Adams and Mr. East are still working on the second malfunctioning printer. Mr. Adams says, "You know I have a trick. What I do with my Radio Shack computer is just to turn it on its side and hit it. Maybe that will work here . . ." They turn it on its side and give it a whack as one of them holds the tension on the paper feed. The machine now works.

The knowledge needed to make minor hardware repairs, to be able to distinguish which problems one can fix and which require outside help, and to operate specific pieces of software is only part of what is needed to utilize computers effectively in teaching. Equally or even more crucial is knowledge about the software available in one's subject area. At least two separate issues arose in this regard at Whitmore. First, teachers need to have a mechanism available for locating software that suits their needs; second, they need some way to evaluate it. Although a few teachers, like Mr. East, were well aware of various information sources about educational software, most teachers were not. Furthermore, since the level of information available was so low, teachers were frequently, even generally, disappointed in the software they did get. Thus a vicious circle occurred, with teachers believing that little good quality software was available, deciding to take a chance and order something, and then having their low expectations confirmed.

"Overload" of Knowledgeable Teachers

Becker's (1984) work suggests that individual teachers very often play a major role in providing the impetus for a school's obtaining instructional computers. Although Becker reports a recent trend for administrators to become more involved in this process than was the case several years ago, individual teachers still play a crucial role in the implementation stage with regard to issues such as deciding what software will be purchased and providing informal training for other teachers.

Such was certainly the case at Whitmore. For example, from all reports, if Ms. Prentiss had not volunteered to supervise the gifted computer lab at lunchtime as her "duty" period instead of monitoring the halls, the gifted computer lab would not have been used on a daily basis. However, observations and interviews at Whitmore revealed a fundamental problem: Few teachers had any substantial knowledge about computers and no formal mechanisms were available for helping interested teachers utilize computer facilities. Thus increased usage of computers within the school meant increased burdens on teachers like Mr. Adams, Mr. East, and Ms. Prentiss. Yet aside from giving Ms. Prentiss "credit" for a duty period, no formal arrangement was ever made to respond to this situation. This put knowledgeable teachers in a position of conflict when colleagues or students requested help, because that help had to be taken from time that was either their own personal time or from parts of the day that were officially allocated to other more traditional uses, namely, teaching or preparing to teach.

> *Interviewer:* What are the one or two major impediments to greater usage of the computers?
>
> *Ms. Prentiss:* The time for *one* person to coordinate the use of the room. . . . What I didn't realize when we started is that a teacher who doesn't have complete control of the class. . . . and know everything about the machines . . . could cause so much damage. They walk out of the room and who's got to deal with it? You know who . . . It's selfish, but I didn't bargain to . . . I often give 2 periods a day and lots of extra time . . . What happened one time that other teacher asked (to use the room) I had to teach her class —make up the dittos (about using the computers). Hey, I don't want word to get out I'm doing this. Then every. . . . teacher . . .

Thus present organizational arrangements stand in the way of optimizing computer utilization. New ways of meeting the lack of knowledge problem previously discussed must be found that do not place too heavy a burden on the teachers who have some interest and expertise. A number of potential solutions are quite obvious and involve familiar mechanisms, such as release time for knowledgeable teachers. However, both thought and research need to be devoted to exploring whether more imaginative solutions might be more effective and to delineating the consequences of the various possible solutions to this problem.

Attitudinal Barriers

We indicated earlier that a widespread lack of knowledge about computers appears to be a major impediment to their utilization. Further, we have recommended research designed to see how this lack of knowledge about both hardware and software can be remedied. Yet such remedies will be effective only if teachers *want* to use computers in their work so that they take advantage of an environment that facilitates this desire. Such may not be the case. At Whitmore, there was evidence of many teachers' indifference to or even resistance to the idea of using computers in their teaching. Our work at Whitmore suggested the importance of learning more about teachers' attitudes to computer usage and the determinants of such attitudes. First we will briefly discuss indicators of indifference and resistance. Then we will proceed to indicate the apparent basis for such feelings.

One reasonable indicator of teachers' enthusiasm for computers is their level of interest in utilizing them or in learning about how to utilize them. Both appeared quite low, as the following excerpts from interviews suggest. The first excerpt is from an interview with Ms. Prentiss:

> *Interviewer:* How have other teachers responded to your efforts to make computers available [by opening the gifted computer lab at lunchtime]?
>
> *Ms. Prentiss:* Generally surprise. I'm a bit of an anomaly. Not that many women do these things and . . . just "Why?" "Why?" It's one thing if you can do *your own* work on it, but why would you (do the extra work)? . . . I've never been so isolated from teachers as I have this year. I don't even eat with them . . . So they . . . think I'm weird because I want to socialize with kids.

We also had a chance to interview Mr. Carter, a member of the mathematics department, about the utilization of the computer tutor:

> *Interviewer:* Have you had a chance to see the computer tutor which Mr. Adams and Mr. Brice . . . [are using]?
>
> *Mr. Carter:* Yeah, but I'm not too fond with computers. I don't want any parts [sic] with computers. I'm the old-fashioned type. I don't want to learn anything new. Maybe that's my fault. I should go into learning computers. I had enough of computer training at Waterford University, but I don't know. I just—after so many

years, you build up a file on your subjects . . . For me to go into teaching computers . . . I would have to start all over. I would have to actually sit down and
work everything out, and it would require a lot more work on my part to run a
class the way I want it run. At this point in time I suppose everybody gets lazy
and . . . I just don't want to do it . . . I'm doing what I'm doing. Don't want to
change.

We have already discussed one possible source of the teachers' general lack
of interest in computers—the threat that the process of learning to use them
poses to teachers' sense of competence. Another related possible source of
this attitude is the perception that computers pose a threat to the teachers'
autonomy and to predictability in teaching. Specifically, the possibility of
"bugs" in the software or of hardware failures means the teacher may need
outside assistance. In addition, a teacher using computers can not be assured
of being able to have students cover the material planned for a specific day.
An excerpt from our field notes illustrates this problem.

> Peter can't get his computer started properly. He looks around and calls out
> "Miss" to Ms. Lee, a staff member on the computer-tutor project. She doesn't
> hear him. He twists all the way around in his chair and calls out quietly, "Some-
> body," but nobody replies. (Mr. Adams and both of the computer-tutor staff
> members are working with other students). Peter looks at me and rolls his eyes
> with an exaspertated expression on his face. He sits passively for a while . . .
> (then) he calls out, "Excuse me, help!" Ms. Lee . . . goes over to him. (Peter
> works independently for about 15 minutes after getting started 10 minutes or so
> after most students). Towards the end of class, when Peter gets up and puts on
> his jacket, I say to him, "How's it going?" He replies, "The hardware is broken
> . . . On the first day we couldn't get it on. Yesterday it wouldn't let me off, and
> today, it broke!" (Later) I asked Ms. Lee what the problem was. She replied,
> "The computer just froze out . . ."

One of the computer-tutor staff remarked during an interview:

> We found the machines behaved wonderfully (in our lab) . . . We get them in
> the schools and we were getting some really strange errors . . . It affected the
> atmosphere and the kids . . . There was *A LOT* of machine breakdown in school
> which frustrated me and it frustrated Mr. Adams and it frustrated the kids. And
> the worst problem was, they weren't the kind of breakdowns that you could
> duplicate. The Xerox man would come and run diagnostics (and not be able to
> find the problem) . . . It was real tough.

Of course, one would expect an unusually high incidence of such problems
during the field testing of new software, but this kind of problem was by no
means unique to the field-test site, as indicated by the field notes on page
18. This concern about the unpredictability of computers and the problems
that this may pose for teachers may well have been reinforced by the fact
that Whitmore, like many schools, was in a transition period between using
manual and computerized means for a variety of clerical and administrative
functions, most especially recording and reporting grades. This transition

was far from smooth and it created a great deal of extra work and annoyance for the teachers. In fact, as a consequence of problems with the new computerized system teachers had to turn in their grades a week earlier than usual and the beginning of summer school was delayed so that marginal students could learn whether they had failed a class before the beginning of summer classes. Such a context was hardly conducive to encouraging teachers to use computers in their classes.

A final factor that seemed to contribute to resistance to utilizing computers was concern about the cost of the machines coupled with a fear that they might somehow be used to replace teachers. Such a concern is hardly surprising, especially with regard to the intelligent geometry tutor. Certainly those involved with the development of that software could foresee a day when the tutor could perform many if not most of a teacher's academic functions.

Interviewer: Do you think the tutor could be developed to the point where it and the student will form a self-sufficient teaching and learning unit?

Mr. Lawton (a member of the computer-tutor development team): I can foresee that . . . You wouldn't necessarily need a human. It's not a particular goal right now, but it's certainly a very viable possibility.

Because they were well aware that the idea of replacing teachers was an extremely volatile issue, the computer-tutor staff took great pains to emphasize that the tutor's goal was to *help* teachers not to *replace* them. However, some teachers found this hard to believe, or at the very least questioned whether expending such sums on a teaching *tool* made sense. The 10 machines utilized in the field test of the computer-tutor cost a total of half a million dollars. Teachers used to using thirty-dollar textbooks and pieces of chalk as teaching tools not surprisingly wondered at the cost-effectiveness of this approach. Few were aware that by the end of the field-test period the tutor was able to operate on the Apple MacIntosh rather than on the $50,000 Dandy Tigers which were used in the field test.

Logistical and Practical Impediments to Computer Usage

Becker's (1986) research shows quite clearly that computers tend to be utilized more when they are grouped in laboratories than when they are spread around in individual classrooms. Thus it is far from unusual to see recommendations that computers be grouped in this way. We would suggest that considerably more thought needs to be given to the question of how computers should be distributed and what the consequences of these various arrangements are. Our research suggested that somewhat different barriers to usage were associated with these two different arrangements. Specifically, when one or just a few computers were available to a class serious organizational and management issues arose, which teachers often just avoided by failing to use the machines. In fact, the computers in the gifted lab had originally been given to individual teachers for classroom use. However, when it became clear that the machines were not being used (indeed some machines had reportedly never even been turned on during their first year at Whitmore) they were collected into a central location.

The existence of a laboratory created its own problems that inhibited usage. First, teachers are very attached to their classrooms, the arenas in which they have the most power and autonomy (Lortie, 1976; Schofield, 1982). Utilizing a lab generally requires some coordination, thus again undermining a teacher's autonomy. Security arrangements, although necessary, heighten barriers by requiring greater coordination and by emphasizing that the room is not the teacher's own. Using a lab takes the teacher off his or her "turf." This sense of going to foreign territory is well-illustrated by the fact that one teacher spontaneously likened taking his class to the computer lab to "taking a field trip inside the school." Finally, it is at least possible that grouping rather than dispersing computers undercuts involvement by removing the machines from most teachers' presence. While it is clear that many teachers let computers sit idle in their rooms, it seems at least possible that others would be encouraged by the easy availability of the resource to consider its possibilities.

Barriers to Recruiting and Retaining Qualified Teachers

The observation that we are facing a shortage of well-qualified mathematics teachers has recently received a great deal of attention, which is hardly surprising in light of the fact that in 1981 over forty states reported a shortage of such teachers (Romberg, 1984). The reasons for this situation are numerous and beyond the scope of this paper. However, we would suggest that many of those very forces that lead to the present shortage will operate even more strongly on teachers with computer skills than on those without. For example, pay differentials are often cited as one reason why teachers leave the classroom. Under most present systems of remuneration, classroom teachers with computer skills are paid no more than those without. Yet such skills are clearly a valuable commodity in the labor market. Thus teachers with such skills may leave teaching in greater numbers or get so involved in after-hours consulting activities that their attention is diverted from their primary job. Although most teachers want to be treated as professionals, their job is not accorded a great deal of status. In sharp contrast, in our society the mystique that surrounds technology lends glamour and status to many computer-related jobs. Thus, those math teachers with especially well-developed computer skills may find it more rewarding to adopt a professional identity relating to computers rather than to teaching. This phenomenon is illustrated by field notes on a conversation between a teacher at Whitmore and a member of our research team.

Mr. Davidson then mentioned (to an observer) that he does a lot of outside consulting on computers. Recently when he was talking with a banker, the banker asked what Mr. Davidson's occupation was. Mr. Davidson replied, "I'm in computers." The banker asked if he was a programmer and Mr. Davidson said he replied, "No, I'm a consultant and a teacher." Mr. Davidson then went on to talk about how he had helped this banker to learn to use a personal computer saying enthusiastically, "I think I have a new client!"

New strategies are clearly needed to provide enough mathematics teachers who are capable of preparing students to function in the coming century. Some such strategies have already been suggested (Romberg, 1984). Others may need to be developed. It is important that we understand not only the extent to which these various approaches are successful in recruiting and retaining personnel, but also which strategies yield what kinds of individuals and what the consequences of these are for teaching and learning.

Differential Access to Computer Resources

Previous research on computer usage in education has raised the question of differential access to computers by different kinds of students (Becker, 1983; Sheingold & Endreweit, 1983). A particularly important research area for the near future is how to make sure that mathematics education is effective for female and minority group students. This issue is important not only as a matter of equity but because population trends make it clear that a smaller and smaller proportion of the students in the United States will be white males (Hodgkinson, 1985).

Traditionally mathematics has been seen as a discipline in which males, especially white males, excel. Computer science, too, has typically been seen as a white male domain.[2] The question then arises of whether utilizing computers to teach mathematics will pose an extra barrier to female and minority group students' learning in mathematics. Our study suggests that this may sometimes be the case. For example, it is probably more than coincidental that the two students who consistently expressed hostility toward or displeasure with the computer-tutor were black females.

As I enter the room, Allen, a white boy from the preceding class, is still working at his computer. He's staring at the screen intently. Mr. Adams says, "Come on Allen! Come on Allen! Come on Allen! You'll be late!" Allen says, in a pleading voice, "Please just let me finish! Let me finish!" Mr. Adams say reluctantly, "Well, okay, another minute or two and that's it!" Annie (black) is working . . . Mr. Adams says jocularly to Ms. Lee, "Did you get that program hooked up? . . . The one that sends a jolt of electricity through her (meaning Annie's) chair when she swears at the computer." Ms. Lee laughs . . . Kelly (a black girl) is trying to get started. She sits for about three minutes (waiting for the computer to present her with a problem). She looks very sulky with her chin on her hand and a pout on her face . . . Ms. Lee says jokingly, "You know, I think what happens is that you type in your name and it sees who you are and says, 'No! No! No!'" Kelly is not amused. Ms. Lee . . . types in some things . . . She says in a hopeful voice, "Third time is a charm," but nothing happens. Kelly now looks almost triumphant. A smile keeps twitching on her lips as . . . nothing appears on the screen

. . . Annie calls out loudly and angrily to the computer, *"What you doin'?"* Giggles run around the room. . . (Later) Kelly is sitting at her terminal doing nothing . . . Ms. Lee says, "Are you totally and completely lost?" Kelly says quietly, "Yeah." Ms. Lee says, "Well then you can ask for system select" (an option which asks the computer to show you the next step). Kelly does but looks puzzled. Ms. Lee says, "It gave you the first step. What rule applies?" Kelly's face is screwed up in a frown. It's hard to tell whether she feels like yelling or crying, but she is clearly full of suppressed emotion. Her foot is tapping rapidly under the table. She says, *"I don't* understand this problem," in a strained voice. Mr. Adams comes over and says soothingly, "What's the problem?" Kelly mutters something I can't hear . . . Mr. Adams strides to his desk, picks up a pad of paper and a pencil and brings it back. He draws a diagram and begins to ask questions. Kelly responds correctly . . .

Although computer science has traditionally been seen as a male domain there are now certain applications of computers, such as word processing, that are often seen as fitting readily with female sex roles. One important challenge for research will be to find ways of incorporating computers into the mathematics curriculum in a way that facilitates mathematics interest and achievement on the part of females and minorities as much as, or even more than, on the part of white males.

Motivators of Utilization

Many forces such as those outlined above conspire to keep teachers from using computers in instruction. It seems logical to argue that two things need to occur before the situation changes markedly. First, we need to learn how to overcome these barriers. Second, we need to learn more about incentives to usage, for even in the absence of major barriers change seems very unlikely to occur without the presence of positive forces leading to it.

At Whitmore High School, three major factors seemed to lie behind the acquisition and utilization of microcomputers. First, there were the relatively rare cases in which a teacher saw some real, very positive instructional purpose to using computers and made a major investment of time and energy so that students could benefit from them. The three clearest cases of this were Mr. Adams, who was very enthusiastic about the possible benefits of the computer tutor, Ms. Prentiss, who supervised the gifted lab, and Mr. Edwards who actually wrote two small grant proposals to acquire money for a modest computer setup for his special education classes.

Second, some teachers, notably Mr. Adams and Mr. East, just plain enjoyed computers. Their own personal enthusiasm for the machines seemed to spill over into a desire to help students learn about them. Although overall their influence was extremely constructive, there did appear to be some disadvantages to having a high proportion of the teachers (outside of computer science) who used computers being motivated to some extent by the sheer love of "hacking." First, as indicated earlier, there was such a gulf in knowledge between people like Mr. East and the rest of the faculty that others sometimes felt intimidated to ask for or unable to understand the information they sought. Second, student-teacher interactions ostensibly designed to teach mathematics or some other substantive area using computers sometimes got transformed into information sessions on hardware, computer programming, or the like, topics which these teachers found extremely interesting. This transformation did not occur when Mr. Adams used the computer-based geometry tutor with his class, but was evident in some of his trips to the gifted computer lab with the same geometry classes.

A third very obvious motivation for the acquisition of computers at Whitmore was a desire to impress the public, especially parents of school-age children. Mr. East candidly discussed the situation in an interview.

> It's [having a computer lab] something you can brag about to parents . . . We're in direct competition with private schools and Mr. Miller, the vice-principals, and the counselors romance the parents at the beginning of ninth grade. "You sure want to send your students here . . . Let me show you what's going on . . ." They [visit] the room downstairs showing them the marvelous new machines . . . which many private schools simply cannot afford.

Interestingly, access to computers was also seen as a badge of status by many within the school. Although Mr. Edwards, the special education teacher used his computer to very good effect, he also felt it served a valuable public relations function for him and his students within the school.

> It's a motivational thing, if nothing else, for me to have it in a special ed class so that mainstream kids passing by as well as colleagues [see it]. There's still a mystique about it . . . They think you're some kind of whiz . . . and all you're doing is punching out stupid discs . . . It gives you a little ego trip. Your colleagues recognize that you're doing something innovative . . .

Unfortunately, to the extent that public relations is a motivation for acquiring computers, the machines serve their purpose at least tolerably well sitting unused. Thus while public relations concerns help to place computers in the school they do not necessarily lead to constructive use of the machines.

The Impact of Computer Usage on Classroom Functioning

The major part of this paper has been devoted to discussing factors which appeared to inhibit computer usage for instructional purposes. However, a second major class of issues that requires further research is the impact that computers have on teaching. Although a relatively small amount of research has been done on this topic, there is evidence that the impact of computers is less consistent than one might expect; that rather than shaping the classroom in highly predictable and clear-cut ways, computer usage is shaped by the context in which it occurs (Sheingold, Kane, & Endreweit, 1983). It seems clear that there is no necessary set of consequences that follow from computer usage, since these machines can be used in so many different ways and to such different ends. Yet our observations of the computer-based geometry tutor at Whitmore suggested that teachers must deal with a number of important issues when using computer-based instruction. How each of these issues is handled may ultimately determine whether and how classrooms are influenced by computer usage. Space limitations make it impossible to do more than to suggest the areas where we believe change that has important educational implications may well occur. As will become apparent, we urge attention both to the changes and adjustments that teachers consciously make when they use computers and to the unintended side effects of computer usage.

Change from Whole-Group to Individualized Instruction

One of the more obvious changes that often accompanies usage of any substantial number of computers in a classroom is a shift from whole-group to individualized or at least small-group instructional techniques. Such a change quite clearly calls on different skills in a teacher. For example, in the high school setting a teacher's prowess as a lecturer becomes less important, while a teacher's ability to respond effectively to individuals becomes more crucial. However, a great many less obvious but potentially important consequences of this shift need to be explored so that we can understand the extent to which they occur as well as their consequences. We will illustrate this point by discussing changes which appeared to occur in the classes using the computer-based tutor.

Changes in Amount and Type of Attention Given to Students at Varying Achievement Levels

Although students using the computer-tutor were tracked into classes of three different ability levels, there was still considerable heterogeneity within classes, especially in the regular classes. Mr. Adams was, of course, very aware of these differences. His clear tendency when working problems at the chalkboard in the whole-class instruction was to call on the more advanced students as previous research has suggested is often the case (Bossert, 1979). This both raised the probability of a correct answer and saved considerable time. In addition, it saved the slower students embarrassment, as can be seen in the following excerpt from our field notes. It also meant that they received less attention from Mr. Adams and often had the answers provided before completing a problem successfully.

> Iris (one of the better students) says heatedly, "That's unfair. You always call on Tom for extra credit! . . ." Mr. Adams doesn't answer her complaint directly. Instead he assigns another problem and says, "Iris will choose who answers this time. This is extra credit." Annie and Peter finish first. They have their hands up. Mr. Adams says, "Okay. Choose." Iris says, "You put your hands down. I want to call on one of those." (She points to where Darlene and Kit, who are clearly the slowest students in this class, are seated). Both girls have their heads bent over their papers, still working. Darlene says to Mr. Adams, "Can I ask a ques-

tion?" Mr. Adams goes over to where she is working and answers it. She continues to work. Iris says, "Are you ready, Kit?" Mr. Adams says, "The bell is going to ring any minute and no one will get credit." He continues sarcastically, "You'd be a good teacher. It's your class. Time is going." Iris goes up to the front of the room. Mr. Adams, in a voice close to a shout, says, *"Would You Please Call On Someone!"* . . . Iris hesitates and then calls on Darlene. Darlene gives 7 angles correctly but gets the 8th one wrong. Peter volunteers the correct answer and Mr. Adams lets him show how he got the answer. He says, "This is the one you all missed so I want you to watch it . . ." (The bell rings). Iris says to Mr. Adams, "Sometimes it's not having the right [answer]—it's having a chance. If you give them a chance . . ." Mr. Adams interrupts heatedly, "Here are people having difficulty. You . . . focus all of the attention on them. Isn't that embarrassing? It puts them in a corner." Iris counters, "Okay, okay, but why don't you ever call on them?" Mr. Adams replies, "You need to learn something about people. They get it wrong. They make bad subtraction errors. You're different. You know where it's at . . ."

When using the computer tutor, in contrast, the slower students often received considerably more attention than the brighter ones. In fact, it was not uncommon for the observer to estimate that Darlene and Kit received 4 or 5 times as much attention from Mr. Adams as did the more advanced students. Such attention was not likely to be embarrasing because students were often unaware of exactly with whom Mr. Adams was working. In addition since students could proceed at their own pace, Mr. Adams' working with the slower students did not impede the rest of the students as much as under a more traditional whole-class method of instruction. Further research is clearly warranted to explore the issue of whether and how using computers changes the teacher's attention to students of various ability levels and whether this in turn influences student learning.

Changes in Traditional Evaluation Practices

The arrival of the computer tutors created a problem with regard to how to grade students. Before their arrival, Mr. Adams used a very traditional point grading system with a certain number of points allocated for homework, tests, and the like. However, since one of the major advantages of the computer-tutor was that it allowed students to proceed at their own pace, grading everyone according to the same standard of accomplishment no longer seemed so appropriate.

Interviewer: Has the introduction of the tutors changed the basis on which you assign grades?

Mr. Adams: This is a problem! Oh my God, yes, how do I grade them? . . . I've had to develop a policy . . . (Let's say) when they came in and started on the tutor they had a grade of C. If they came in everyday and worked everyday and made a legitimate effort, they'd go up to a B . . . A half-assed effort, they'd go down to a D. If they came and didn't give a damn at all they'd go down to an E . . .

Interviewer: So really effort is the main thing now.

Mr. Adams: . . . Effort will mean a lot more . . . It had to. See, I'll be honest with you . . . I just don't buy effort. It doesn't mean much to me, it doesn't . . . A college is going to look at that grade . . . so I can't give a B for effort. The grade has got to reflect what they know. Let's face it, these things [tutors] are going to be here for one report period . . . It's not going to change their grade for the year by any real wild difference.

Interestingly, the chairman of the math department remarked to our project staff that he could not evaluate Mr. Adams very well since the class was run so differently from ordinary ones and different skills were needed. Thus the utilization of the tutor raised questions about teacher evaluation as well as about student evaluation.

There is no reason to assume that computer usage will automatically change grading practices. Yet research by others studying very different kinds of computer usage also suggests that using computer technology has important implications for assessment (Hawkins & Sheingold, 1986). Thus the issue seems worthy of further exploration.

SUMMARY AND CONCLUSIONS

We would argue that research on computers and mathematics instruction needs to be grounded in a conception of why and how computers can be effectively used. Such a conception must be based on an analysis both of what students need to learn and of the special capabilities and characteristics of computers. Since software differs so dramatically in purpose, scope, and content, researchers trying to understand computers' impact on teaching may find it necessary to employ a strategy like that utilized in the study discussed here; that is, studying a particular kind of computer usage. Although in some ways such an approach is quite limiting, it has significant advantages as long as a wise choice is made about the kind of usage studied. Two important choice criteria, which are not necessarily correlated, are extensiveness of usage and instructional promise. Study of the former lets one assess what is happening while study of the latter lets one assess what could happen.

A major thrust of any research agenda on using computers in mathematics teaching must be attention to both barriers and incentives to usage. Usage of computers in teaching mathematics requires many types of change. *Effective* usage of these resources undoubtedly will require even more change. We have outlined a few of the important barriers to change suggested by our own research. Undoubtedly there are many other equally deserving of attention. Unless, and until, barriers are reduced and incentives are increased, there is little reason to think that computers will be utilized to anything like their full potential. Issues of barriers and incentives to change must be addressed at many levels: economic, organizational, and psychological, to name just a few. Research must not only isolate the most important barriers and incentives but also focus on how to change the present situation, in which barriers to effective usage often outweigh incentives.

A focus on barriers and incentives implies an emphasis on utlization, specifically, a desire to see greater utilization of computers in mathematics teaching. However, utilization per se, is a sterile, even potentially dangerous, goal. The reason for desiring utilization, increased learning of valued material, must be constantly kept in mind. Thus a second major focus for future research concerns the consequences, both intended and unintended, of computer usage. Reflection on the issues we have suggested here as worthy of study raises an important point. The mere study of outcomes is not sufficient. Rather we would suggest attention to the processes that mediate those outcomes. A focus on processes and mediating variables has two major advantages. First, it is more illuminating from a scientific perspective than straight input-output research. Second, it is likely to be a much more fertile source of ideas on how best to equip teachers to handle the new demands and to take advantage of the new opportunities which computers present.

NOTES

[1] To be more precise, we asked all computer-tutor students and their parents to agree to the students' participation in interviews. A similarly sized sample of control students were also invited to participate in the study. On the whole, cooperation was excellent, with 90% of the computer-tutor students and 82% of the control students who were selected for interviews participating.

[2] Asian-American students often seem to do unusually well in mathematics (Brand, 1987), thus creating one notable exception to this admittedly broad generalization.

REFERENCES

Anderson, J. (1984). *Proposal to the Carnegie Corporation to support demonstration and development of a geometry tutor*. Unpublished proposal, Carnegie-Mellon University, Pittsburgh, PA.

Anderson, J. R., Boyle, C. P., Farrell, R., & Reiser, B. J. (1984). *Cognitive principles in the design of computer tutors*. Unpublished paper, Carnegie-Mellon University, Pittsburgh, PA.

Anderson, J. R., Boyle, C. P., & Yost, G. (1985). *The geometry tutor*. Paper presented at International Joint Conference on Artificial Intelligence. Los Angeles.

Becker, H. J. (1983). School uses of microcomputers. *Reports from a National Survey, Issue 3* (pp. 1-8). Center for Social Organization of Schools, The Johns Hopkins University, Baltimore, MD.

Becker, H. J. (1984). School uses of microcomputers. *Reports from a National Survey, Issue 4* (pp. 1-5). Center for Social Organization of Schools, The Johns Hopkins University, Baltimore, MD.

Becker, H. J. (1986). Instructional uses of school computers. *Reports from the 1985 National Survey, Issue 1* (pp. 1-9). Center for Social Organization of Schools, The Johns Hopklns University, Baltimore, MD.

Bogdan R. C., & Biklen, S. K. (1982). *Qualitative research for education: An introduction to theory and methods*. New York: Allyn & Bacon.

Bossert, S. T. (1979). *Tasks and social relationships in classrooms: A study of instructional organizaion and its consequences*. New York: Cambridge University Press.

Brand, D. (1987, August 31). The new whiz kids: Why Asian-Americans are doing so well and what it costs them. *Time*, 42-51.

Committee on Research in Mathematics, Science, and Technology Education, Commission on Behavioral and Social Sciences and Education, (1985). *Mathematics, Science, and Technology Education: A research agenda.*

Washington, DC: National Academy Press.

Goetz, J. P., & LeCompte, M. D. (1984). *Ethnography and qualitative design in educational research.* New York: Academic Press.

Hawkins, J., & Sheingold, K. (1986). The beginning of a story: Computers and the organization of learning in classroooms. In J. A. Culbertson & L. L. Cunningham (Eds.), *Microcomputers and Education.* Chicago: The University of Chicago Press.

Hodgkinson, H. L. (1985). *All one system: Demographics of education, kindergarten, through graduate school.* Washington, DC: The Institute for Educational Leadership, Inc.

Lesgold, A. M., & Lesgold, S. B. (1984). *Classroom computers and state curriculum policy.* Unpublished manuscript, Learning Research and Development Center, University of Pittsburg, Pittsburg, PA.

Lortie, D. C. (1975). *School teacher: A sociological study.* Chicago: The University of Chicago Press.

McKnight, C., Crosswhite, F., Dossey, J., Kifer, E., Swafford, J., Travers, K., & Cooney, T. (1987). *The underachieving curriculum: Assessing U.S. school mathematics from an international perspective.* Champaign, IL: Stipes Publishing Company.

National Council of Teachers of Mathematics (1980). *An agenda for action: Recommendations for school mathematics in the 1980s.* Reston, VA: National Council of Teachers of Mathematics.

Olson, S. (1976). *Ideas and data: Process and practice of social research.* Homewood, IL: The Dorsey Press.

Reichardt, C. S., & Cook, T. D. (1979). Beyond qualitative versus quantitative methods. In T. D. Cook & C. S. Reichardt (Eds.), *Qualitative and quantitative methods in evaluation research* (pp. 7-32). Beverly Hills, CA: Sage.

Romberg, T. A. (1984). *School mathematics: Options for the 1990s.*

Chairman's Report of a Conference, Vol. 1. Washington, DC: U.S. Government Printing Office.

Schofield, J. W. (1982). *Black and white in school: Trust, tension or tolerance?* New York: Praeger.

Schofield, J. W. (1985). *The impact of an intelligent computer-based tutor on classroom processes: An ethnographic study.* Unpublished paper, University of Pittsburgh, Pittsburgh, PA.

Schofield, J. W., & Anderson, K. (1987). Combining quantitative and qualitative components of research on ethnic identity and intergroup relations. In J. S. Phinney & M. J. Rotheram (Eds.), *Children's ethnic socialization: Pluralism and development* (pp. 252-273). Beverly Hills, CA: Sage.

Sheingold, K., Hawkins, J., & Char, C. (1984). "I'm the thinkest, you're the typist": The interaction of technology and the social life of classrooms. *Journal of Social Issues, 40*(3), 49-62.

Sheingold, K., Kane, J. H., & Endreweit, M. E. (1983). Microcomputer use in schools: Developing a research agenda. *Harvard Educational Review, 53*(4), 412-432.

Smith, L. M., & Geoffrey, W. (1968). *The complexities of an urban classroom.* New York: Holt, Rinehart and Winston.

Taylor, W. D., & Johnson, J. B. (1986). Some issues in the application of technology to education. In J. A. Culbertson & L. L. Cunningham (Eds.), *Microcomputers and education* (pp. 216-233). Chicago: The University of Chicago Press.

Walker, D. F. (1984). Promise, potential and pragmatism: Computers in high school. *Institute for Research in Educational Finance and Governance Policy Notes, 5*(3), 3-4.

Walker, D. F. (1986). Computers and the curriculum. In J. A. Culbertson, & L. L. Cunnlngham (Eds.), *Microcomputers and education* (pp. 22-40). Chicago: The University of Chicago Press.

Willis, B. H., Thomas, S. N., & Hoppe, M. H. (1985). *Technology and learning: Changing minds in a changing world*. Southeastern Regional Council for Educational Improvement, Schooling and Technology Series, 4.

The research on which this paper is based was funded by the Office of Naval Research (ONR), contract number N00 14-85-K-0664. However, all opinions expressed herein are solely those of the authors and no endorsement by ONR is implied or intended.

Cross Cultural Studies of Mathematics Teaching and Learning Recent Findings and New Directions

James W. Stigler

The University of Chicago

Michelle Perry

The University of Michigan

"I'm sick and tired of hearing about the Japanese."
> —*Comment by the Mathematics Coordinator for an affluent school district in Cook County, Illinois, as told to a cross-cultural researcher.*

The comment above is not unusual, but then neither is the opposite expression of wholehearted enthusiasm for gaining understanding of how it is that Japanese children so dramatically outperform their U.S. counterparts in tests of mathematics achievement. What is it about Asian mathematical superiority that produces such divergent responses?

If we pause to take stock of what we know about Asian mathematical performance, both responses make a good deal of sense. On the one hand, the achievement differences between Asian and U.S. children are astounding. A recent study reported by Stevenson, Lee, and Stigler (1986) compared a representative sample of fifth-grade classrooms in Sendai, Japan and Taipei, Taiwan with a sample of classrooms from the Minneapolis metropolitan area. On a test of mathematics achievement the highest-scoring U.S. classroom did not perform as well as the lowest-scoring Japanese classroom and outperformed only one of the twenty Chinese classrooms. This is a whopping difference, which leads many to be enthusiastic about learning from the Japanese.

On the other hand, although the achievement differences are large and often reported in the media, the possible mechanisms that underlie these differences are not well studied, and remain largely unknown. Teachers in the United States are often "hit over the head" with achievement differences across countries. Yet the meaning of achievement rarely has moved beyond a single score on a single test, and explanations of the differences rarely move beyond the gross level of "country." Under these circumstances, just knowing that the Japanese students are outperforming U.S. students can lead to frustration.

If we know only that the differences exist, but do not have a good description of the cultural and educational context that surrounds such differences, it is no wonder we get sick and tired of hearing about the Japanese. What are we to make of such differences if we do not know more about what kinds of mathematical knowledge Asian children excel in learning? And how can we use such differences to inform our own practice of teaching mathematics if we know very little about how mathematics is taught in Asian schools?

The purpose of this paper is to begin to supply the research context that must surround cross-cultural studies of mathematics teaching and learning if such studies are to be useful for U.S. teachers and concerned citizens in general. We can learn from the Japanese, and from other countries as well, but only if we collect relevant information and interpret that information in a sensible way. As one author has put it (White, 1987), Japan provides us not with a blueprint for the education of U. S. children, but rather with a mirror that we can use to examine ourselves. The point of this paper is that cross-cultural studies of mathematics teaching and learning in general have the potential to provide us with just such a mirror.

The Cultural Context of Mathematics Teaching and Learning

Of all the things children learn in school, mathematics would seem to be the one subject least affected by culture. After all, numbers are numbers, and basic mathematical operations should function the same across cultures.

Although saying that "numbers are numbers" is clearly a generalization, it is the relatively transcultural nature of mathematics that makes it especially interesting for cross-cultural study. With more verbal domains of learning, it is difficult to disentangle skill from context. Not only does the context of learning vary across cultures, but so does the content of what is being learned. The nature of mathematics, on the other hand, provides the researcher with a point of commonality on which to build a comparison. The relative similarity of the body of mathematical knowledge across cultures thus makes the role of culture itself more, rather than less, prominent in understanding how children acquire mathematical knowledge, precisely because the content remains the same while the cultures vary.

In most nations around the world, mathematics is taught in school, where we find materials, activities, a teacher and a learner. (This is not to imply that mathematics is taught exclusively in school, but our primary focus here is upon mathematics learning that occurs in school.) The cultural context of mathematics learning can be described through the tools, traditions, beliefs, attitudes, and practices that support the learning of school mathematics. A brief description of these aspects of culture as they have an impact upon mathematics learning in school will be provided here.

Cultural differences are found to some degree in the tools and methods children are provided with for the performance of basic mathematical operations. Some of the most basic mathematical skills have their roots in verbal skills and thus may be influenced by cross-linguistic differences in the ways in which numbers are represented verbally. For example, recent research by Miller and Stigler (in press) has shown that learning to count is easier for children who speak Chinese than it is for English speakers, due primarily to the more consistent structure of the list of count words in Chinese. In addition, the Chinese number names of the digits 0 to 9 can be pronounced more quickly than can the English number names, thus leading to the fact that Chinese speakers can remember longer strings of numbers in a standard digit span task than can speakers of English (Stigler, Lee, & Stevenson, 1986).

Apart from the language itself, different cultures provide different representational tools for such tasks as counting and computation. Saxe (1981) and Lancy (1983) have described how peoples in Papua New Guinea use body parts as a means for counting objects, and Zaslavsky (1973) provides a rich description of indigenous African counting systems. Many Asian cultures teach children to add, subtract, multiply and divide using an abacus, or sometimes even a visual image of an abacus (Stigler, 1984; Stigler, Chalip, & Miller, 1986).

Beyond these culture-specific technologies on which some mathematical skills are constructed, there are large cultural differences in the beliefs held by parents, teachers, and children about the nature of mathematics learning. These beliefs can be organized into three broad categories: beliefs about what is *possible*, (i.e., what children are able to learn about mathematics at different ages); beliefs about what is *desirable* (i.e., what children should learn); and beliefs about what is the best *method* for teaching mathematics (i.e., how children should be taught).

Beliefs about what is possible. Cultural differences in conceptions of the possible have been found in numerous recent studies. For example, in work comparing Japanese, Chinese, and U.S. parents, it has been found that U.S. mothers are significantly more likely than Japanese mothers to believe that innate ability (as opposed to effort) underlies children's success in mathematics (Stevenson, Lee, & Stigler, 1986; Lee, Ichikawa, & Stevenson, in press). Clearly, if innate ability is believed to determine success in mathematics, then there are always limits on what one could expect a particular child to achieve in school mathematics.

Differing beliefs about what is possible also are expressed in the design of school mathematics curricula. U.S. textbooks limit children in various ways, probably because of beliefs about what is possible to teach children of different ages. For example, U.S. elementary textbooks introduce large numbers at a slower pace than do Japanese, Chinese, or Soviet textbooks, and delay the introduction of regrouping in addition and subtraction considerably longer than do books in the other countries (Fuson, Stigler, & Bartsch, in press). Although we tend to believe that it is best to introduce small numbers before large numbers, and addition without regrouping before addition with regrouping, the fact that Asian children do so well using curricula not based on such beliefs suggests that our beliefs might profitably be questioned.

Beliefs about what is desirable. Beliefs about what children should learn also differ across cultures. Although U.S. elementary schools spend far less time teaching mathematics than do schools in Japan or Taiwan (Stigler, Lee, & Stevenson, 1987), parents in the U.S. still believe that reading, not mathematics, needs more emphasis in the curriculum than it currently receives (Stevenson, Lee, & Stigler, 1986). Within the mathematics curriculum, there also may be large cross-cultural differences in beliefs about which topics should be stressed. For example, some have debated whether school mathematics should be oriented toward problem solving in the real world, or whether it should be more purely mathematical (cf. Lave, Smith, & Butler, in press).

Beliefs about the best methods for teaching mathematics. Finally, there are many cultural beliefs that relate to the proper method of instruction. Two interesting domains in this regard are beliefs about the nature of understanding, and about the role of concrete experience in children's learning. Those in the U.S., particularly with respect to mathematics, tend to assume that understanding is equivalent to sudden insight. With mathematics, one often hears teachers tell children that they "either know it or they don't," implying that mathematics problems can either be solved quickly or not at all (cf. Schoenfeld, 1985). In Japan and China, understanding is conceived of as a more gradual process, where the more one struggles the more one comes to understand. Perhaps for this reason, one sees teachers in Japan and China pose more difficult problems, sometimes so difficult that the children will probably not be able to solve them within a single class period. U.S. teachers, by contrast, tend to pose problems that will reinforce the idea that mathematics problems should be solvable in a single, insightful motion.

Related to beliefs about the nature of understanding are beliefs about the role of concrete experience. U.S. teachers tend to believe that young children need concrete experiences in order to understand, and even that concrete experiences will automatically lead to understanding. These beliefs are questioned when one observes a Japanese classroom, as discussed later in this paper. Japanese teachers apparently believe that even young children can understand abstraction and that concrete experience must be accompanied by reflection in order for understanding to follow.

Scope of this Paper

Having laid out the broader sense in which culture intrudes on the learning of mathematics, we will now narrow the scope somewhat for the remainder of the paper. While we believe that mathematics learning is always influenced by wider cultural meanings, we will restrict the rest of this paper to a discussion of cross-cultural studies of the learning of mathematics in school, in particular in modern industrialized countries where schooling is nearly universal. Thus, the focus will be on curriculum, teaching, and achievement. First, we will lay out the potential contributions that could be made by cross-cultural studies of mathematics teaching and learning in school, and then briefly review the relatively few studies that have been done. Then, we will outline the design of the University of Michigan study of mathematics learning in Japan, Taiwan, and the United States and present preliminary results from observations of first-grade mathematics classes collected as part of that study.

WHAT CAN BE LEARNED FROM CROSS-CULTURAL COMPARISON?

Before reviewing the cross-cultural studies that have been done or that are in progress, it is useful to pause first to consider both what we might expect to learn from such studies and what we cannot expect to learn. The most important thing that we should not expect to learn is the causal mechanisms that underlie high or low mathematical achievement in different cultures. For one thing, something as complex as student learning will be caused by multiple factors, and it is difficult, if not impossible, to ascertain the way in which numerous factors combine to determine achievement. In addition, the limits of comparative studies are well known (cf. Campbell & Stanley, 1963). In the absence of experimental control, the most that we can hope for are relatively well-justified hints about what is really going on.

There are, however, some very distinct advantages of cross-cultural studies. By looking closely at the teaching and learning of mathematics within the schools of different countries we can expand our own beliefs about what is possible to expect in our own culture. These new possibilities may be in terms of level of achievement, possibilities for pacing or sequencing, or new tools and methods that we were not previously aware of. Further, by observing what succeeds and what fails in other cultures, we can begin to formulate new possibilities for teaching in our own culture; we can use the natural variation that occurs across cultures to guide us in new directions, without having to explore each of those directions anew for ourselves.

When Japanese students peform as highly as they do on tests of mathematics achievement, new possibilities are opened up for the level of achievement we might think possible for U.S. students to attain. We may not have thought that it was possible or valuable for elementary school students to be formally exposed to probability and statistics. Yet, when we see Japanese children successfully learning topics such as these, we can imagine the possibility for U.S. children. In this case, we can observe behavior in Japan that cannot be seen in the U.S. This sort of variation across cultures is so striking because it allows us to realize that different possibilities exist, not only in how something is taught, but also in whether or not something can be taught. The same can be said for the pacing and sequencing of the topics already common to Japanese and U.S. curricula. New possibilities can be imagined for how much time is necessary to teach a given topic; or we can question our assumption that topic A really must be taught before, rather than after, topic B.

New techniques for teaching can also be discovered by observing mathematics teachers in other countries. For example, simple addition and subtraction facts are generally taught by U.S. teachers either using straight memorization or, less frequently, by counting on (Fuson, 1982). In China and Japan a different technique is widespread, namely the decomposition and recomposition of numbers into groups of ten. Japanese teachers have developed TILE representations for teaching this technique (Hatano, 1982), and Chinese teachers use similar materials (e.g., strips of paper divided into ten squares with plastic flowers for filling the squares). The point is that here is a technique that is new to U.S. teachers, but that has been well-tested by Asian teachers. Cross-cultural studies allow us to benefit from the experience of a much wider range of teachers and to discover new techniques that may be usable in our own culture.

Cross cultural comparison also leads researchers and educators to a more explicit understanding of their own implicit theories about how children learn mathematics. Without comparison, we tend not to question our own traditional teaching practices and we may not even be aware of the choices we have made in constructing the educational process. For example, by observing classrooms in other cultures we can find that teachers place a large emphasis upon the whole class working together, but when observing classrooms in our own culture we find that teachers place an emphasis upon working individually. From this sort of comparative work we can gain insight into our own beliefs about how learning occurs.

In general, cross-cultural comparison allows us to observe wider variability, both in teaching and achievement, than can ordinarily be observed within a single culture. In addition, aspects of mathematics teaching that are covariant in one culture may be unconfounded in another, thus making it possible to question culture-based assumptions about the way in which two variables must be related.

A BRIEF LOOK AT PREVIOUS CROSS-CULTURAL RESEARCH

Several cross-cultural studies of mathematics learning exist. The most ambitious of these have been the studies conducted by the International Association for the Evaluation of Education Achievement (IEA) (e.g., Husen, 1967; McKnight, Crosswhite, Dossey, Kifer, Swafford, Travers, & Cooney, 1987; Travers, Crosswhite, Dossey, Swafford, McKnight, & Cooney, 1985). The first IEA study (Husen, 1967), carried out in 1964, measured achievement in various mathematical topics in each of 12 different countries, at two grade levels: 8th grade and 12th grade. The second IEA study (known as SIMS, for Second International Mathematics Study) compared 17 countries in the 8th-grade component and 12 in the 12th-grade component. Both the IEA and SIMS studies measured 8th grade students' abilities to solve arithmetic, algebra, geometry, statistics, and measurement problems; and measured 12th grade students' abilities to solve algebra, geometry, elementary functions and calculus, probability and statistics, sets and relation, and number-system problems.

A major finding from both the IEA and SIMS studies was that the United States did not perform as well as had been expected. Among the countries studied, the United States consistently performed at or below the median level in each of the topic areas tested. Countries that consistently outperformed the U.S. included Japan, Hong Kong, and Belgium.

An example of the U.S.'s poor performance—and a cause for many U.S. educators to feel alarmed—can be found by examining results of the arithmetic computation subtest. The SIMS documents that, unlike most other participating countries, the U.S. is still emphasizing instruction in arithmetic in the 8th grade; however, performance in arithmetic for U.S. students is below the average for all participating countries.

The SIMS went beyond the first IEA in that it attempted to explore some of the underlying causes for the notable differences in achievement. Although the SIMS investigated many possible causes for cross-cultural differences in mathematics achievement, many of the factors were not found to relate to student achievement. For example, neither class size nor years of teacher training were related to country differences in student achievement. The authors of the SIMS report suggest that the lack of a consistent relationship between amount of formal teacher training and student achievement may be due to the fact that teachers in all countries are required to undergo some sort of formal training and are exposed to colleagues, in-service training, and other forms of continued education.

Besides ruling out uninfluential factors, the SIMS was successful at discovering two likely contributors to student mathematics achievement: the status granted to teachers in different countries (and its concomitant responsibilities and potential benefits) and the curriculum that was presented to the students. The SIMS noted that Japanese teachers are accorded far higher status than U.S. teachers and that Japanese students achieved more. The authors of the SIMS report (McKnight, et al., 1987) hypothesized that teachers' status might influence student achievement in the following way: The higher a teacher's status, the less time that she or he will be required to spend on noninstructional activities (e.g., time spent at required administrative meetings). A teacher who has fewer demands on her time will have more time to prepare for class and, thus, will presumably teach better classes. In fact, Japanese teachers were found to spend less time on administrative tasks and more time preparing for classes than U.S. teachers.

The curriculum is another variable that has been identified as influential in students' learning. The SIMS documented what they term the "implemented curriculum" by having teachers note which items on the achievement test teachers had taught to their classes. The SIMS confirms that students who are not taught calculus, for example, cannot solve calculus problems. Certainly, if students in one country have performed well or poorly on a particular type of problem, then it is useful to know whether that type of problem was or was not introduced to students.

Although the IEA and SIMS studies have covered a lot of ground in describing cross-cultural differences in mathematics achievement, other studies have also made contributions to this area. For example, Harnisch, Walberg, Tsai, Sato, & Fryans (1985) completed a large comparison of Japanese and U.S. high school students. Harnisch et al. found that Japanese students outperformed U.S. students and investigated some of the possible causes of this difference. They examined several possible contributors to achievement differences, including which mathematics courses students have completed and home background variables. The dependent variable used in this comparison was a test that measured student skill in several topic areas (including algebra, geometry, etc.). Based on several analyses, Harnisch et al., concluded, as did the SIMS, that which courses students take plays a major role in achievement differences, whereas other factors play a less influential role in mathematics achievement.

Other studies have looked at curriculum apart from achievement. For example, Fuson (Fuson, Stigler, & Bartsch, in press), and Stigler (Stigler, Fuson, Ham, & Kim, 1986) have investigated the grade levels at which addition and subtraction topics are introduced in the primary grades. Fuson et al. reported that relatively difficult addition and subtraction topics (e.g., subtraction problems that require borrowing from a zero in the subtrahend) are introduced very late in U.S. mathematics texts compared to when these topics are introduced in Soviet, Taiwanese, Mainland Chinese, and Japanese texts. Stigler et al. reported comparable findings for the introduction of addition and subtraction word problems in U.S. and Soviet texts; Soviet texts present many different types of word problems, whereas U.S. texts present only a few types of word problems. Achievement was not measured in these studies and, thus, they cannot tell us that a certain component within the curriculum has a direct influence upon a specific area of achievement. However, a more precise measurement of the curriculum, such as that provided by Fuson's and Stigler's analyses, may be valuable in exploring the exact nature of achievement differences, because previous work has already provided evidence for the link between curriculum and achievement (cf. Harnisch et al., 1985; McKnight et al., 1987).

Most previous cross-cultural work has focussed on the achievement of high school students. Only rarely have researchers attempted to trace the roots of these later achievement differences by studying younger children.

One study that has looked at younger children is reported by Song and Ginsburg (in press). This study measured mathematical skills of Korean and U.S. children at several ages, from children enrolled in day-care centers through the third grade. Song and Ginsburg found that, through the first grade, U.S. children showed higher levels of performance than Korean children, but this advantage disappeared by the second and third grades. Song and Ginsburg's results suggest that schooling may play an important role in the development of observed cross-cultural differences in mathematics achievement. However, more work is needed before we can pinpoint what in particular about schooling is causing differences in children's achievement.

We have briefly noted the range of studies that have investigated cross-cultural differences in mathematics achievement. Most of the studies have demonstrated country differences in achievement, but most have not gone far beyond the documentation of differences. What is needed is to begin to break down both the independent variable of "country" and the dependent variable of "achievement" into more fine-grained units that can deepen our understanding of both the underlying mathematical knowledge and the external cultural factors that may underlie the achievement differences.

The SIMS work on curriculum begins to analyze the components of the country effect. However, there are a great many other aspects of culture that need to be studied: teachers' and students' beliefs about mathematics learning, the role of mathematics in the everyday culture, and classroom processes involved in the teaching of mathematics are all factors that remain relatively unstudied. "Achievement," similarly, needs to be broken down into more meaningful units: Are Japanese students better at all aspects of mathematics, or only at some aspects? Are there other outcomes we might want to study as well as achievement, such as estimation skills, visual problem-solving skills, and children's future goals relating to mathematics?

The University of Michigan studies were designed to fill in some of these gaps in the cross-cultural literature. Both the first and second Michigan studies have focused on mathematics in the elementary school, an age level that has been sorely neglected in the previous research. In addition, the focus has been on those areas that have been most neglected by previous work: beliefs, attitudes, and classroom processes.

AN OVERVIEW OF THE MICHIGAN STUDIES: BACKGROUND AND DESIGN

The University of Michigan studies comparing academic achievement of children in Japan, Taiwan, and the United States began in 1978, and are presently being continued. There have been two major waves of data collection: the first in 1979-80 and the second in 1985-86. Some background on the goals of the two studies may be helpful.[1]

The First Study

The first study began with an emphasis on reading, in particular, on the role of orthography in learning to read and in the generation of reading disabilities. Although the results of the reading investigation (Stevenson, Stigler, Lucker, & Lee, 1982) proved interesting, the most striking cross-cultural differences emerged in the area of mathematics (Stigler, Lee, Lucker, & Stevenson, 1982; Stevenson, Lee, & Stigler, 1986). Although differences in mathematics achievement had been noted previously between Asian and U.S. children (e.g., Husen, 1967), never had such differences been found among elementary school children, and certainly not as early as the first grade.

The first study included only a single, individually administered test of mathematics achievement, one that had been carefully constructed for the purposes of our study. The test items focused primarily on computation (and also geometry for fifth-grade students). The test was administered to a sample of first- and fifth-grade students in Sendai, Japan; Taipei, Taiwan; and the Minneapolis metropolitan area. In each city, 10 representative schools were selected, and within each school 2 first- and 2 fifth-grade classrooms participated. Twelve children in each classroom were tested, yielding a final sample of 480 children in each of the 3 locations. The results of the test are presented in Figure 1.

Many other kinds of data were collected in the first study, including interviews with parents, cognitive testing, and classroom observations. However, it soon became clear that, given the large cross-cultural differ-

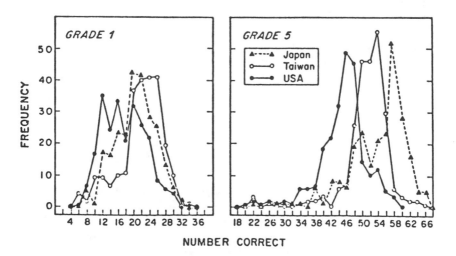

Fig. 1. Frequency distribution of scores obtained by children in Grades 1 and 5 on the mathematics test in Japan, Taiwan, and the United States. (From: Stigler, Lee, Lucker, & Stevenson, 1982.)

ences in mathematics achievement, it would be well worth our while to design a subsequent study directed specifically at understanding the cross-cultural differences in mathematics learning. We designed the study with two major goals in mind. First, we wanted to go beyond the single outcome measure produced by a standard achievement test and find out more about what kinds of mathematical knowledge Japanese, Chinese and U.S. children had acquired. Second, we wanted to probe more directly, through interviews and observations, into the teaching, curriculum, and societal context of mathematics achievement.

The Second Study

The mathematics study was again conducted in Sendai, Japan and Taipei, Taiwan. In the United States, however, we decided to use the Chicago metropolitan area instead of Minneapolis. Minneapolis is a city with a very low percentage of minority and of non-English-speaking children. The Chicago area is far more diverse in population, and thus more representative, in many respects, of mainstream United States.

In each of the two Asian cities, 10 schools were selected to participate in the study. In the Chicago area, 20 schools were chosen to represent the urban and suburban areas that make up Cook County. The decision to include twice as many schools in Cook County as in the other two locations was based on the far greater diversity of children in Cook County. Within our Chicago sample of schools, we included public and private schools; upper, middle, and lower socioeconomic status neighborhoods; predominantly Black, White, Hispanic, or ethnically-mixed schools; and urban and suburban environments. The schools were chosen in collaboration with local educational authorities to be representative of the range of schools found throughout each metropolitan area.

Near the beginning of the school year—October in Taiwan and the United States, and May in Japan—each classroom was visited by a team of testers who administered a group test of arithmetic computation and a test of reading comprehension. A total of 5524 children were tested in the three sites, spread across 160 classrooms.

Within each classroom, a subsample of 3 boys and 3 girls was randomly selected for further study. The subsample consisted of 480 children in Cook County and 240 children in each of the other two locations. The additional information on each child in the subsample was collected towards the end of the school year in which group testing had been conducted. Each child in the subsample received approximately two additional hours of individually administered mathematics tests; their mothers, teachers, and school principals were interviewed; they themselves were interviewed; and their classrooms were observed during four separate mathematics classes.

The purpose of the additional testing was to broaden our understanding of what specific knowledge differences underlie Asian superiority in mathematics achievement. There were eight mathematics tests in all, administered in two separate individual testing sessions. All of the tests were especially constructed for this study, and were judged fair by a team of researchers representing each of the cultures being studied. The tests included word problems, operations, visualization, graphing, mental calculation, number concepts, estimation, and mental image transformation.

The interviews were designed to gather information about the family and educational context in which the children's mathematical skills were developing. The "Mother Interview" included basic socioeconomic and educational data about the family and about adult and sibling involvement in the child's learning of mathematics (e.g., "How often do you help your child with his/her mathematics homework?"). In the "Teacher Interview" we obtained information about the teacher's training, about his or her beliefs on learning mathematics, and about the curriculum used (including who decides what gets taught and the teacher's opinion about that curriculum). Information regarding the entire school (including age and condition of the physical plant, number of teachers, responsibilities of teachers and students beyond teaching and learning, etc.) was obtained through an interview with the school principal.

Results from the tests and interviews are now being analyzed, and will be published soon. We will devote the remainder of this paper to analyses of the classroom observations, since this information is most directly informative about mathematics teaching in these three cultures.

MATHEMATICS TEACHING IN JAPAN, TAIWAN, AND THE UNITED STATES: SOME PRELIMINARY ANALYSES

In this section we will present findings from classroom observations conducted during both the first Michigan study, and the second, mathematics study. A full report of observations in mathematics classrooms from the first study is published elsewhere (Stigler, Lee, & Stevenson, in press). However, the method employed there, and some of the more important findings, will be reviewed to provide background for interpreting results from the second study. Methods used in the second study will be described, and very preliminary analyses of the first-grade classrooms will be reported. (A full report will follow.)

Observational Method: First Study

Each of the 120 first- and fifth-grade classrooms was visited 40 times over a two- to four-week period. The visits were scheduled to yield a stratified random sample of time across the school day and school week, thus making it possible to estimate the amount or percentage of time that was devoted to various activities. (A full description of the method can be found in Stigler, Lee, & Stevenson, 1987).

Each visit lasted about an hour, and included time for separate observations of teachers and of individual students. The procedure was to observe the target, either teacher or child, for 10 seconds, and then to spend the next 10 seconds coding the presence or absence of a checklist of categories. This procedure was repeated according to a predefined sequence that counterbalanced order of observation across the teacher and the 12 randomly chosen target students in each class. Across the two- to four-week observation period, each of the 12 children in each classroom was observed for about 33 minutes (not including coding time), and each teacher was observed for about 120 minutes.

The student coding system included 30 categories, although coding was eased somewhat by the fact that many of the categories were mutually exclusive. Various aspects of the classroom were coded from the target child's point of view, including the following: Was the class engaged in academic activities, or in transition between activities? What subject matter was being taught? How was the classroom organized and who was the leader of the child's activity? And what kinds of on- and off-task behaviors was the child engaged in.

The teacher coding system contained only 19 categories. These categories noted who the teacher was working with; what kinds of teaching behaviors the teacher was engaged in; and what kinds of feedback the teacher was offering to the students.

Specific categories from the student and teacher coding schemes will be introduced as the results are presented (although only some results from the first study will be recapitulated here). Our major emphasis will be on describing observations of first-grade classrooms completed as part of the second study.

Observational Method: Second Study

Observations for the mathematics study differed in two important ways from observations conducted in the first study. First, only mathematics classes were observed, making it impossible to compare mathematics teaching with the teaching of other subject matters. Second, in addition to an objective coding system, narrative descriptions were recorded in each class.

Each of the 160 classrooms in the mathematics study was visited four separate times over a one- to two-week period, yielding a total of 640 observations across the three locations. Observers, who were local residents of each city, arrived just before teachers began the daily mathematics lesson and observed until the mathematics class was over. Observers worked in pairs, with one observer doing the category coding and the other observer doing the narrative descriptions. The objective category coding was similar to that used in the first study and has not yet been analyzed. Thus, only analyses of the narrative observations will be presented here.

The narrative observers were instructed to write down as much as they could about what was transpiring during the class. Their goal was to record the on-going flow of behaviors and to include descriptions of all supporting materials (e.g., what was written on the blackboard, how many children were working on which problem, etc.). Observers were instructed to use a set of abbreviations common across the three countries, which enabled the observers to spend more time recording details of the class. The observers also noted, with marks in the margin, when one minute had elapsed. These minute markers were included so that we would be able to estimate the duration of various activities.

Time, Organization, and Disorganization: Findings From the First Study

The results of the first observational study served mainly to differentiate classrooms in the United States, on one hand, from classrooms in Japan and Taiwan, on the other. In the first study, very few differences emerged between Chinese and Japanese classrooms. In some respects, one only has to visit one Chinese or Japanese classroom to see vast differences between Asian and U.S. elementary school classrooms. Class size is a major difference: while the classrooms in our Minneapolis sample averaged 22 students in the first grade and 24 students in the fifth grade, the classrooms in Taipei averaged 45 and 48 students at the two grade levels, and those in Sendai, 39 at both grade levels. Most Asian classrooms are arranged with desks in rows facing the teacher, while U.S. classrooms often have desks arranged in groups.

The two dimensions on which the cultures varied most obviously were in time spent on the teaching and learning of mathematics and in the level of organization in the classroom.

Time. Children in Japan and Taiwan spend significantly more time in school than do children in the United States, and this ultimately translates into Japanese and Taiwanese children spending significantly more time learning mathematics. School is in session 240 days per year in both Japan and Taiwan, compared to only 180 days per year in the United States. Although first-graders in all three cities that we studied spent about 30 hours per week in school, fifth-graders in Sendai spent 37 hours a week in school, those in Taipei, 44 hours, and those in Minneapolis, still only 30 hours.

Although we observed only during academic classes, and not during such periods as lunch, gym, recess, or assemblies, the students, nevertheless, were not always engaged in academic activities. In first grade, U.S., Chinese, and Japanese children spent 69.8%, 85.1%, and 79.2% of the time, respectively, engaged in academic activities. At the fifth grade the corresponding percentages were 64.5%, 91.5%, and 87.4%. At both grade levels, Chinese and Japanese children spent a much higher percentage of their time engaged in academic activities than did U.S. children. Furthermore, although the percentage of time spent in academic activities increased between first and fifth grade for the Asian children, the percentage actually declined slightly across grade levels for the U.S. children.

Our observers recorded the percentage of time devoted to different subject matters. The majority of time in all three cultures was devoted to either reading/language arts or to mathematics. Although the total percentage of time devoted to either one of these two subject matters was similar across the three cultures, the way in which time was apportioned between the two varied significantly by culture. As is apparent in Figure 2, U.S. teachers at both grade levels devoted more time to reading/language arts and less time to mathematics than did Chinese and Japanese teachers. By the fifth grade, both Chinese and Japanese teachers spent approximately equal amounts of time teaching mathematics and reading. U.S. teachers, by contrast, spent almost three times as much time on reading as they did on mathematics.

Calculations based on the hours per week spent in school, the percentage of time spent in academic activities, and the percentage of time those aca-

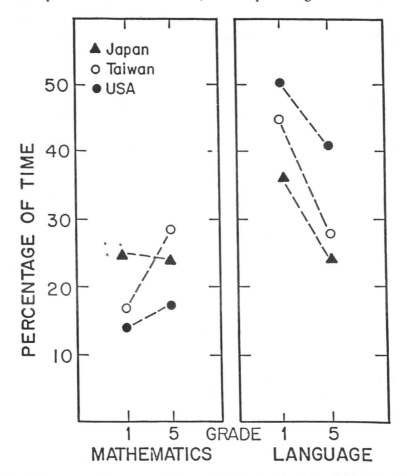

Fig. 2. Percentage of time spent teaching mathematics in Japan, Taiwan, and United States first- and fifth-grade classrooms. (From Stigler, Lee, & Stevenson, in press.)

demic activities were mathematics, versus reading/language arts, allow us to estimate the number of hours each week children in the three cultures spend working on the different subject matters. The results of these calculations are presented in Table 1. The cross-cultural differences in the number of hours devoted to mathematics instruction are large—sufficiently large, in fact, that they could go a long way toward explaining the cross-cultural differences in mathematics achievement.

Level of Organization. The second dimension that differentiated U.S. mathematics classrooms from those in Japan and Taiwan was the level of organization apparent in the classroom. Classrooms in Japan and Taiwan were highly organized and orderly; those in the United States more disorganized and disorderly. These differences were indicated in the coding system in various ways.

Three sets of categories dealt with the way in which the classroom was organized during mathematics instruction. In one set, observers coded whether the target child was working as an individual, as part of a small group, or as part of the whole class. The second set coded similar information, but from the point of view of the teacher: Was the teacher working with the whole class, a small group, an individual, or no one at the time of the observation? In the third set of categories, observers coded who was the leader of the activity in which the target child was engaged: the teacher or no one.

The results of both student and teacher observations regarding the unit of organization (whole class, group, or individual) are presented in Figure 3. Japanese and Chinese students spent the vast majority of their time

TABLE 1

Number of Hours Each Week Spent in

Language Arts and Mathematics

	Country		
	U.S.A.	Taiwan	Japan
Mathematics			
Grade 1	2.9	3.9	6.0
Grade 5	3.4	11.4	7.6
Language Arts			
Grade 1	10.6	10.5	8.8
Grade 5	8.2	11.2	7.8

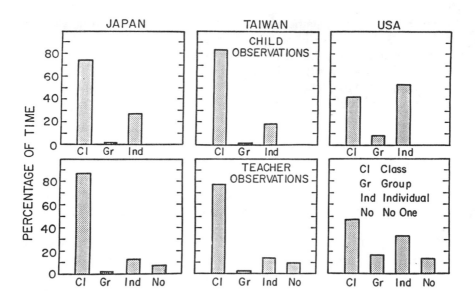

Fig. 3. Percentage of time (a) children spent working as a member of the whole class, a small group, or as an individual (top panel); and (b) teachers spent working with the whole class, a small group, an individual student, or no one (bottom panel). (From Stigler, Lee, & Stevenson, in press.)

working, watching, and listening together as a class, and were rarely divided into smaller groups. U.S. children, by contrast, spent the majority of their time working on their own, and a smaller amount of time working in activities as members of the whole class. The same picture emerges when teachers are observed (the lower panel of Figure 3). U.S. teachers spent more time working with individuals and less time working with the whole class than did Chinese or Japanese teachers. In addition, U.S. teachers were coded in mathematics classes as working with no students 13% of the total time, as opposed to only 6% of the total time for Japanese teachers and 9% of the total time for Chinese teachers.

The counterpart to these findings is displayed in Figure 4, where we see what percentage of the total time in mathematics classes students were part of a teacher-led activity, and what percentage they were part of an activity with no leader. In Taiwan, the teacher was the leader of the children's activities 90% of the time, as opposed to 74% of the time in Japan and only 46% of the time in the United States. No one was leading the student's activity 9% of the time in Taiwan, 26% of time in Japan, and 51% of the time in the United States.

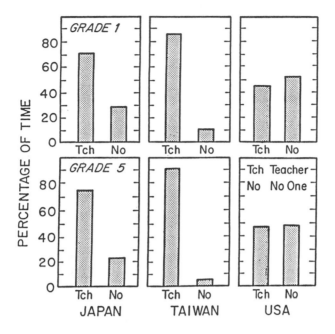

Fig. 4. Percentage of time students spent in activity led by the teacher and by no one. (From Stigler, Lee, & Stevenson, 1987.)

Taken together, these findings indicate that classrooms in the Asian cultures are more hierarchically organized, with the teacher directing her energies to the whole class, and with students more often working under the direct supervision of the teacher. Because of these differences in organization, U.S. students experience being taught by the teacher a much smaller percentage of time than do the Asian students, even though U.S. classes contain roughly half the number of students.

In addition to the relative disorganization that characterizes U.S. classrooms, there is a relative disorderliness as well. This disorderliness was picked up in our coding system by a set of categories for coding the incidence of inappropriate or off-task student behaviors. If the target child was not doing what the teacher expected him or her to do, he or she was judged as being off-task. Two categories of off-task behaviors were distinguished: those behaviors involving inappropriate peer interaction, and those the target child engaged in alone. In addition, we coded whether or not the target child was out of his or her seat. The results from these observations are presented in Figure 5.

There were large cross-cultural differences in the overall percentage of

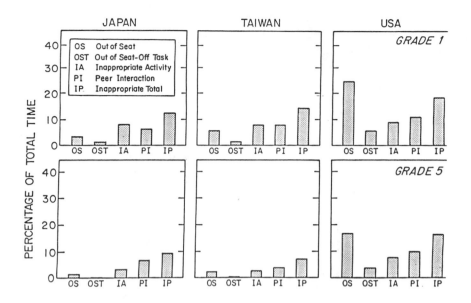

Fig. 5. Percentage of time students in the three countries were coded as engaged in various off-task activities. (From Stigler, Lee, & Stevenson, in press.)

time students spent engaged in inappropriate, off-task activities. Across both grade levels, during mathematics class U.S. students were off-task 17% of the time, as opposed to only 10% of the time for Chinese and Japanese students.

U.S. students were coded as being out of their seats during mathematics classes 21% of the time, whereas Chinese and Japanese children were out of their seats 4% and 2% of the time, respectively. Of course, students being out of their seats does not necessarily imply that they are off-task, particularly in U.S. classrooms. However, if we look at the percentage of time students were *both* out of their seats *and* off-task, the American percentage was 5 times as high as that in the other two countries (5% versus less than 1% in Japan and Taiwan).

Coherence and Reflectivity: Preliminary Ideas from First-Grade Narrative Observations

The data derived from the first observational study is informative in some respects; indeed, we get a clear picture of the frequency with which classrooms are organized in various ways, and we get basic information about how time is spent by students in the three countries. However, even though the observations reported above were made in mathematics classes, we learned very little about how mathematics is actually taught in the three cultures. The narrative observations collected in the current study provide us with richly detailed information concerning what happens in mathematics classes in Sendai, Taipei, and Chicago. We are only beginning our analyses of these data, and will present some early hunches based only on observations of first-grade classrooms. To anticipate what these hunches are, we find two important dimensions along which the classrooms in the three cultures vary.

The first is coherence, from the child's point of view. We will argue, for now at least, that both Chinese and Japanese classrooms provide more opportunities than U.S. classrooms for the students to construct a coherent representation of the sequence of events that make up a typical mathematics class and to understand the goals of the activities in which they are engaged. The second dimension is one that ranges from an emphasis on performance, on the one hand, to an emphasis on reflection and verbalization, on the other. On this dimension, the classrooms in Taiwan tend toward the performance end, and those in Japan toward reflectivity. U.S. classrooms appear confused in this regard, and accomplish neither goal well.

We will first present some information regarding the way in which the narrative observations are being coded. Then we will turn to a fuller account of how we have arrived at the dimensions of coherence and reflectivity as significant and interesting ways to differentiate between classrooms in the three cultures.

Coding. The rich nature of the data gathered in narrative observations exacts its cost later when it is necessary to code the data. We were faced with 640 different narrative descriptions of mathematics classes, in three different languages. Not all observations were of equal quality; in every location, some observers recorded more detail than others, and some were more consistent in their use of abbreviations. How were we to code and summarize the data into a form that would be useful in characterizing cross-cultural differences in mathematics teaching?

We first convened a group of bilingual and trilingual coders to simply read all of the observations. These coders spent weeks reading observations and summarizing their contents in English for the other members of the research group. In addition, a subset of the observations were translated verbatim into English. In this way, we developed a feel for the range of situations we would have to code and some intuitions about cross-cultural differences that would be worth coding. We decided that we wanted some predefined categories that we could apply to the observations, but that we also wanted to preserve a great deal of the detail so that it could be further analyzed later. The coding system we constructed represented a balance of all of these needs.

We decided to begin the coding process by dividing e ich observation into segments, which would be our basic unit of analysis. We found that, as we read through the descriptions of classes, it was relatively easy to divide the class into natural-seeming segments and that we had relatively high agreement amongst group members about where to make the divisions. We gradually developed a more explicit definition of segments by attending to the conditions under which we would say that the segment had changed. A segment was defined as changing if there was a change in *either* topic, materials, or activity.

Topics were globally defined, including categories such as telling time, measurement, or addition facts. Materials included such items as textbooks, worksheets, the chalkboard, or flashcards. Activities, again, were rather molar: examples included seatwork, students solving problems on the chalkboard, or teachers giving explanations. All categories were inductively derived based on our first pass through the data, and we felt that the categories we developed were sufficient to describe the classrooms from the three cultures. The categories were not intended as the full description of the class, but rather as a way of organizing the information into a more useful format.

In addition, an English-language summary was constructed of each segment that would convey in some detail what was going on during the segment. The summary was not intended to be a translation, but rather a briefer recapitulation of the contents of the observation. A great deal of detail still was maintained, however, such as direct quotes from the teachers and students as they participated in the mathematics class. Summaries were written in narrative form so that their contents could be examined later by English speakers not fluent with Chinese or Japanese. The summaries were standardized somewhat by the use of keywords that would serve to alert us to the presence or absence of certain categories in the classroom. For example, whenever a student was observed asking a question to the teacher, the summary would include the standard keyword "S-to-T" so that a computer search for all such situations would be facilitated. Our goal was to make the summaries as consistent as possible in their style and language.

Coherence. After reading the corpus of first-grade observations, all of us in the coding group were struck with the sense that the Chinese and Japanese classes provided more opportunities for the students to construct a coherent account of the sequence of events and activities that make up a mathematics class. In other words, it appeared to us that the Chinese and Japanese classes were in some sense more comprehensible than were the U.S. classes.

The meaning we attached to the term *coherence* is similar to that used in the literature on story comprehension. Of particular relevance is work by

Stein and her colleagues on defining the components of a well-formed story (e.g., Stein, in press; Stein & Policastro, 1984) and work by Trabasso on the role of coherence in story comprehension (e.g., Trabasso & van den Broek, 1985). A well-formed story, which also is the most easily comprehended, consists of a protaganist, a set of goals, and a sequence of events that are causally related to each other and to the eventual realization of the protaganist's goals. An ill-formed story, by contrast, might consist of a simple list of events strung together by phrases such as "and then . . .," but with no explicit reference to the relations among events. The important point is that ill-formed stories are particularly difficult to comprehend, and even more difficult for children to comprehend than for adults. Thus, a certain amount of coherence in input is required if the listener is to be able to construct a coherent representation of the story.

The analogy between a story and a mathematics classroom is not perfect, but it is close enough to be useful for thinking about the process by which children might construct meaning from their experiences in mathematics class. A mathematics class, like a story, consists of sequences of events related to each other and, hopefully, to the goals of the lesson. What we tend to find in the U.S. classroom observations, unfortunately, are sequences of events that go together much like those in an ill-formed story. If it is difficult for adult observers to construct a coherent representation of the events that constitute a first-grade mathematics class, then it surely would be impossible for the average six-year-old to do so.

What are some of the devices employed by Japanese and Chinese teachers to provide more coherence across the events that constitute a mathematics class? One of the major devices we have found is the tendency in Japan and Taiwan to spend an entire 40 minute mathematics class period on the solution of only 1, 2, or 3 problems. A problem thus, in a sense, serves as the protaganist that runs as a single thread through the story, a natural link to tie different segments together.

This devotion of an entire class period to a single problem would seem excessive to U.S. educators. In no class did we observe a U.S. teacher sticking with a single problem for so long, and, indeed, it appears that U.S. teachers value just the opposite approach. In recent research that examined characteristics of expert mathematics teachers in the United States, it was reported that the expert elementary mathematics teacher can get through 40 problems in a single class, whereas the novice teacher may only cover 6 or 7 problems (Leinhardt, 1986; Leinhardt & Greeno, 1986). It would appear that Japanese or Chinese teachers are striving for a different goal. Or, perhaps they are just adapting to a different reality; the value placed on homework in both of these Asian cultures means that repetitive practice can be accomplished at home and class time can be reserved for teaching. U.S. teachers must, especially at the first-grade level, accomplish both purposes during the school day.

It is important to note that this does not mean that Japanese and Chinese classrooms are boring for lack of variety. Variety as indexed by change in segment is approximately equal across the three cultures: The typical first-grade mathematics class in all three cities consists of 5 or 6 segments, each lasting 7 or 8 minutes. What is different is the nature of the changes that occur from one segment to another. While in Japanese and Taiwanese class-rooms segment changes are more often coded because of changes in materials or activities, without a change in topic being taught, in U.S. classrooms the changes are more often coded because of a change in topic being taught (also see Berliner & Tikunoff, 1976). In Japan, only 6.9% of segment changes are marked by changing topic, in Taiwan, 16.1% and in the United States, 24.8%.

Remember that a change in topic does not mean merely a change in problem, but rather a change on the order of, say, starting with measurement and moving to multi-digit addition. For example, one first-grade U.S. class started with a segment on measurement, then proceeded to a segment on simple addition, then to a segment on telling time, and then to another segment on addition. The whole sequence was called "math class" by the teacher, but it is unclear how this sequence would have been interpreted by a child. In this case, it seems that it would be impossible for anyone to construct a coherent account of the whole class.

In other cases, the sequence itself *could* be construed coherently, but U.S. teachers do little to help the child construct a coherent representation. A good example of this kind of situation is provided by the topic of measurement as it is normally taught in first-grade classrooms. Most U.S. textbooks teach fundamental measurement in the following sequence: First they teach children to compare quantities directly, and to say which is longer, wider, and so forth. Next, nonstandard units of measurement are introduced, and children are taught to ascertain, for example, how many paper clips long their pencils are. Finally, students are introduced to the concept of standard units and taught to measure objects in inches or in centimeters. This is a sensible sequence and could conceivably be taught in a coherent manner.

Let us examine the way in which this sequence is implemented in one U.S. classroom in our sample. In the first segment, the teacher has children examine objects (pencils, crayons, paper clips, chalk, etc.) to determine which are longer. The teacher then moves the class to the next segment, and the following quotation begins at the point of transition:

> OK, open your workbooks to page 12. I want you to measure your desk in pencils, find out how many pencils it takes to go across your desk, and write the answer on the line in your workbooks. [Children carry out instructions.] Next see how many paper clips go across your desk, and write that number next to the paper clip in your workbook. [Children continue to follow instructions.] OK, the next line says to use green crayons, but we don't have green crayons so we are

going to use blue crayons. Raise your hand if you don't have a blue crayon. [Teacher takes approximately 10 minutes to pass out blue crayons to students who raise their hands; coded as a transition segment.] Now write the number of blue crayons next to the line that says green crayons. [Teacher then moves on to the third segment.] OK, now take out your centimeter ruler and measure the number of centimeters across your desk, and write the number on the line in your workbooks.

What is fascinating about this particular class is that there is absolutely no marking by the teacher of the transition points—the three segments just follow each other as though there were no transition. There is no discussion of how each exercise is important in providing students with an understanding of measurement: no discussion of why units are important, or why standard units are important; no discussion of the historical development of measurement procedures that could provide more meaning to the sequence of activities; and no discussion of the goals of the class, and how each activity relates to those goals. More time is devoted to making sure students have a blue crayon, which is totally irrelevant to the purpose of the lesson, than to conveying the purpose of the three segments on measurement.

If we put ourselves in the child's position, what is the likelihood that we would construct a coherent, meaningful account of this particular class? Although each of the three measurement activities is intended to provide some new insight within the context of the preceding activity, the rationale behind the sequence is not made clear to the students. As someone has remarked, it is like giving children the punchline without the joke[2]: Ideally, the students would say "I get it!" as each new segment unfolds. But can they "get it" without some explicit reference to the links that tie the segments together? It is highly unlikely, especially for a group of six-year-olds.

In Chinese classrooms, and in Japanese classrooms to an even greater extent, we see teachers providing explicit markers to aid children in inferring the coherence across different segments within a lesson, and across different lessons. Rarely is the logical flow of an Asian first-grade class broken to pursue irrelevant business (such as passing out blue crayons) that may give students the wrong ideas about what is important about mathematics. Transitions often are marked by verbal discussion of the relation between two segments, and classes, especially in Japan, often start with the teacher explaining the goal of the day's class and how the activities relate to the goal.

One Japanese first-grade teacher was quoted as asking this question to a student at the beginning of a mathematics class: "Would you explain the difference between what we learned in the previous lesson, and what you came across in preparing for today's lesson?" To hear a question of this sort posed to a six-year-old would be surprising to most U.S. educators. Perhaps more surprising is that the student was able to answer the question. This kind of interchange highlights the attention paid in Asian classrooms to the students' conscious construction of a coherent account of the classroom experience.

Reflectivity. The second dimension we will discuss is reflectivity: Classrooms can vary in the degree to which they emphasize performance and practice, on one hand, versus reflective thinking and verbalization, on the other. This is an interesting dimension because it appears to differentiate Chinese classrooms from Japanese classrooms, with the Chinese classrooms being more performance oriented and the Japanese classrooms more reflective. U.S. classrooms do not seem to take a definite stand on this dimension. They are not at all reflective, like the Japanese classrooms, nor do they place a consistent emphasis on performance, like the Chinese classrooms. As we will show, at times they attempt both, but end up more often in confusion.

Perhaps a good way to begin to understand this dimension is through looking at verbalization. One index we have of verbalization is through our coding of the incidence of explanations, either by teachers (EXP-T) or by students (EXP-S). EXP-T or EXP-S could be coded as the main activity that characterizes a segment, or as summary keywords describing transitory events embedded within other segments. The incidence of explanations is presented in Table 2.

There are large differences between Japan and the other two cultures in the incidence of verbalization. We have not yet carried out the formal analyses, but it appears from our reading of the observations that verbalization is often used by Japanese teachers as a means of relating different activities to each other and as a means of discussing the principles that underlie different mathematical procedures. The high incidence of verbal explanation in the Japanese classrooms is especially interesting, given that we are observing first-grade classrooms. From the U.S. point of view, actions, not

TABLE 2

Incidence of Explanations in

Japanese, Chinese, and U.S. Classrooms

	Japan	Taiwan	U.S.A.
Percentage of Segments:			
Teacher Explains (EXP-T)	18.0	.3	4.0
Student Explains (EXP-S)	5.0	.3	.1
Percentage of Summaries:			
Containing EXP-T	40.0	13.0	16.0
Containing EXP-S	15.0	10.0	4.0

words, are supposedly more successful means of communicating with a six-year-old.

The Japanese emphasis on reflectivity is further illustrated by comparing statements made by Japanese and Chinese teachers. The Chinese teachers emphasize getting the right answer quickly, whereas Japanese teachers often tell students that the answer is unimportant. Japanese teachers stress the process by which a problem is worked and exhort students to carry out procedures patiently, with care and precision. The Chinese teachers emphasize "do," the Japanese teachers, "think." In fact, the word "think" appears in our Japanese protocols more than twice as frequently as it does in either the Chinese or U.S. protocols.

The Chinese emphasis on speed and on getting the answer is evidenced by the following excerpts from the observations:

- Teacher hands out an April calendar, and instructs students to fill in the missing dates as quickly as they can.

- Teacher runs a competition for "speeded" mental calculation, and writes students' names on the board in order of their speed.

- Teacher evaluates students' blackboard work with a "check" if it was correct *and* fast, and an "X" if it was *either* incorrect or slow.

The Japanese emphasis on reflection and verbal discussion is illustrated in the following excerpts from observations:

- Teacher leads a discussion with students on "which is the best method" for solving a particular problem.

- Teacher directs students' attention to a list of numbers on the blackboard and asks them to look for patterns: "What do you notice here?

- Teacher writes a word problem on the blackboard and tells students the problem is to come up with an equation that can be used to solve the problem. Teacher asks students to choose a partner, and "think about it together in pairs".

- Teacher puts the problem $30 + 60$ on the blackboard, and tells students she wants them to "think about the problem for a whole minute" before beginning to solve it.

- Teacher has one student solve a problem on the blackboard. When the student is finished, he turns to the class and says, "Am I correct?" The class answers, "Not exactly . . .", and then proceeds to correct him.

- The teacher says: "The answer is 41, but that is not as important as the method by which you get it. The crucial thing is the right way to getting the answer."

It is hard to determine where U.S. classrooms fall on this dimension of reflectivity, partly because of the relative disorganization alluded to earlier. It is our impression that U.S. teachers are giving mixed messages to students and confusing them about the goals of mathematics. While Japanese and Chinese teachers are communicating consistent messages about what they view as the most important goal of mathematics instruction (i.e., performance or reflective understanding), U.S. teachers make statements that we find confusing with respect to determining the goals of instruction. U.S. first-grade teachers rarely produced the type of statement that would encourage reflection about mathematics and instead tended to produce statements that were misleading about the goals of mathematics. Here are some examples of statements made by U.S. teachers in our sample:

- On the speed/accuracy issue: "Speed is not as important as neatness."
- "I'm giving you another chance to correct your papers so you don't get any red 'X's on your work."
- Teacher explains: "Let's think about counting by 2's . . . what you do is skip a number, say a number, skip a number, say a number, etc."
- Teacher queries whole class: "What is the rule for subtraction?" Class responds chorally: "The big number goes first."

These are anecdotes, and we must await more careful analysis before making statements about how representative they are of teachers in the three cultures. However, we do not believe we have misled the reader by the particular quotes we have chosen to report. Obviously, there are some superb teachers in our U.S. sample. But the statements we have chosen do, we believe, represent a significant amount of what we see in first-grade classrooms. If we were the students in these classes, what would we construe as the nature and goals of mathematics? This, it seems, is a question worth pursuing.

CONCLUSION

We began this paper with a general discussion of mathematics teaching and learning in its cultural context. We then narrowed our focus to school mathematics and described the kinds of cross-cultural studies that have been done on the teaching and learning of school mathematics. We then launched into a more detailed description of the University of Michigan studies. We ended with some preliminary analyses of narrative observations of first-grade mathematics classrooms in Japan, Taiwan, and the United States, collected as part of the second Michigan study.

An important point, in our opinion, is the amazingly small number, and relatively narrow scope, of cross-cultural studies that have been done. The only major effort aside from the Michigan studies has been the work of the IEA and SIMS. The IEA and SIMS studies have been ground breaking in their analyses of curriculum and achievement, but have not pursued some of the more important cultural factors that surround the teaching of mathematics, nor have they examined student outcomes other than achievement.

What have been particularly lacking, in our opinion, are studies of how mathematics is taught in classrooms in different cultures. It is for this reason that we devoted the greatest part of the paper to reporting analyses of classroom observations from the Michigan studies. As we discussed earlier, there are many aspects of culture that are brought to bear in the teaching and learning of mathematics: beliefs, attitudes, practices, tools, and traditions. There can be no doubt that what happens in the classroom is in some sense a reflection of the wider society in which the classroom exists. Nevertheless, if we want to reform mathematics teaching, it seems that the classroom is a good place to start. Although it is difficult to change what happens in classrooms, it is far more difficult to change broader aspects of the culture.

It is important to emphasize that findings from our observations, or from any other studies that may be done in the future, are not meant as an indictment of U.S. teachers. Indeed, one of the more striking aspects of our findings is the difficult challenge U.S. teachers face each day as they enter their classrooms. Few would want to be in their shoes. We must get beyond the tendency to assign blame if we are to make maximum use of what can be learned from cross-cultural studies of mathematics teaching and learning. There is a great deal that we can learn about ourselves by carefully observing others. We hope others are encouraged to do cross-cultural studies and to deal with the difficult issues of interpretation that inevitably arise. The knowledge that can be gained is worth the difficulty.

NOTES

[1]The Michigan studies have been directed by Professor Harold Stevenson at the Center for Human Growth and Development at the University of Michigan, and conducted in collaboration with numerous colleagues (in addition to the authors of this paper): Shin-ying Lee at the University of Michigan; Chen-chin Hsu at National Taiwan University Medical College; Lian-wen Mao of the Taipei Bureau of Education in Taiwan; and Seiro Kitamura, S. Kimura and T. Kato of Tohoku Fukushi College in Sendai, Japan. The first study was supported by NIMH grants MH 33259 and MH 30567. The second study was supported by the National Science Foundation.

[2]We are indebted to Richard A. Shweder for this analogy.

REFERENCES

Berliner, D. C. & Tikunoff, W. J. (1976). The California beginning teacher evaluation study: Overview of the ethnographic study. *Journal of Teacher Education, 27*(1), 24-30.

Campbell, D. T. & Stanley, J. C. (1963). *Experimental and Quasi- experimental designs for research.* Chicago: Rand McNally.

Fuson, K. C. (1982). An analysis of the counting on procedure in addition. In T. P. Carpenter, J. M. Moser, & T. A. Romberg (Eds.), *Addition and subtraction: A cognitive perspective* (pp. 67-81). Hillsdale, NJ: Erlbaum.

Fuson, K., Stigler, J., & Bartsch, K. (in press). Grade placement of addition and subtraction topics in Japan, Mainland China, the Soviet Union, Taiwan, and the United States. *Journal for Research In Mathematics Education.*

Harnisch, D., Walberg, J., Tsai, S-L, Sato, T., & Fryans, L. (1985). Mathematics productivity in Japan and Illinois. *Evaluation in Education,9*, 277-284.

Hatano, G. (1982). Learning to add and subtract: A Japanese perspective. In T.P. Carpenter, J.M. Moser, & T.A. Romberg (Eds.), *Addition and subtraction: A cognitive perspective* (pp. 211-223). Hillsdale, NJ: Lawrence Erlbaum Associates.

Husen, T. (1967). *International study of achievement in mathematics*. NY: Wiley.

Lancy, D. F. (1983). *Cross-cultural studies in cognition and mathematics*. New York: Academic Press.

Lave, J., Smith, S., & Butler, M. (in press). Problem solving as everyday practice. In E. Silver & R. Charles (Eds.), *Teaching and evaluation of mathematical problem solving*. Reston, VA: National Council of Teachers of Mathematics.

Lee, S., Ichikawa, V., & Stevenson, H. (in press). Beliefs and achievement in mathematics and reading: A cross-national study of Chinese, Japanese, and American children and their mothers. In M. Maehr (Ed.), *Advances in motivation: Vol. 7*. Greenwich, CT: JAI Press.

Leinhardt, G. (1986). Expertise in math teaching. *Educational Leadership*, *43*(7), 28-33.

Leinhardt, G. & Greeno, J.G. (1986). The cognitive skill of teaching. *Journal of Educational Psychology*, *78*(2), 75-95.

McKnight, C., Crosswhite, F., Dossey, J., Kifer, E., Swafford, J., Travers, K., & Cooney, T. (1987). *The underachieving curriculum: Assessing U.S. School mathematics from an international perspective*. Champaign, IL: Stipes.

Miller, K. & Stigler, J. (in press). Counting in Chinese: Cultural variation in a basic cognitive skill. *Cognitive Development*.

Saxe, G. B. (1981). Body parts as numerals: A developmental analysis of numeration among the Oksapmin in Papua New Guinea. *Child Development*, *52, 306-316*.

Schoenfeld, A. H. (1985). Mathematical problem solving. New York: Academic Press.

Song, M. J. & Ginsburg, H. P. (in press). The development of informal and formal mathematical thinking in Korean and U.S. Children. *Child Development*.

Stein, N. L. (in press). The development of children's storytelling skills. In M.B. Franklin & S. Barten (Eds.), *Child language: A book of readings*. New York: Oxford University Press.

Stein, N. L. & Policastro, M. L. (1984). The concept of a story: A comparison between children's and teachers' perspectives. In H. Mandl, N. Stein, & T. Trabasso (Eds.), *Learning and comprehension of text* (pp. 113-155). Hillsdale, NJ: Lawrence Erlbaum Associates.

Stevenson, H. W., Lee, S. Y., & Stigler, J. W. (1986). Mathematics achievement of Chinese, Japanese, and American children. *Science*, *231*, 693-699.

Stevenson, H. W., Stigler, J. W., Lucker, G. W., and Lee, S. Y. (1982). Reading disabilities: The case of Chinese, Japanese, and English. *Child Development*, *3* 1164-1182.

Stigler, J. W. (1984). "Mental Abacus": The effect of abacus training on Chinese children's mental calculation. *Cognitive Psychology*, *16*, 145-176.

Stigler, J., Chalip, L. & Miller, K. (1986). Consequences of skill: The case of abacus training in Taiwan. *American Journal of Education*, *94*(4), 447-479.

Stigler, J., Fuson, K., Ham, M., & Kim, M. (1986). An analysis of addition and subtraction word problems in U.S. and Soviet elementary mathematics textbooks. *Cognition and Instruction*, *3*(3), 153-171.

Stigler, J. W., Lee, S. Y. & Stevenson, H. W. (1986). Digit memory in Chinese and English: Evidence for a temporally limited store. *Cognition*, *23*, 1-20.

Stigler, J. W., Lee, S. Y. & Stevenson, H. W. (in press). Mathematics classrooms in Japan, Taiwan, and the United States. *Child Development*.

Stigler, J. W., Lee, S. Y., Lucker, G. W., & Stevenson, H. W. (1982). Curriculum and achievement in mathematics: A study of elementary school children in Japan, Taiwan, and the United States. *Journal of Educational Psychology*, *74*(3), 315-322.

Trabasso, T. & van den Broek, P. (1985). Causal thinking at story comprehension. *Memory and Language*, *24*, 612-630.

Travers, K. J., Crosswhite, F. J., Dossey, J. A., Swafford, J. O., McKnight, C. C., & Cooney, T. J. (1985). *Second international mathematics study summary report for the United States*. Champaign, IL: Stipes.

White, M. (1987). *The Jananese educational challenge: A commitment to children*. New York: Free Press.

Zaslavsky, C. (1973). *Africa counts*. Boston, MA: Prindel, Weber, & Schmidt.

This paper was prepared for the NSF-NCTM Research Agenda Conference on Effective Mathematics Teaching, Columbia, Missouri, March 1987. The paper was written while the first author was supported by a Spencer Fellowship from the National Academy of Education, and the second author by funds from the Benton Center for Curriculum and Instruction at the University of Chicago. Research described in the paper was funded by NSF grant BNS8409372.

Can Teachers be Professionals?

Thomas A. Romberg
Professor, Curriculum and Instruction
Faculty Associate, Wisconsin Center for Education Research
University of Wisconsin—Madison

The question posed in the title of this paper has been raised because I do not believe that mathematics teachers in U.S. schools are now true professionals, nor can they be, given the current view of the job of teaching held in our society and the constraints on that job in most school systems. In fact, the contradiction that exists between societal expectations for teachers and the limitations that society in turn places on them is at the root of the problem. On the one hand we give teachers the authority to influence children and even bestow a "social identity" on them that frequently channels their entire adult lives (Romberg & Price, 1983). On the other hand, in too many schools the job of teaching has been deskilled to such an extent that it can not be called a professional occupation. As I stated in an earlier paper:

> Timid supervisors, bigoted administrators, and ignorant school boards often inhibit real teaching. A commercially debauched popular culture makes learning disesteemed. The academic curriculum has been mangled by the demands of both reactionaries and liberals. Attention to each student is out of the question, and all the students—the bright, the average, and the dull—are systematically retarded one way or the other, while the teacher's hands are tied. Naturally, the pay is low for the work is hard, useful, and of public concern. (Romberg, 1985, p. 4)

In this paper I describe the conditions that could provide the basis for a different conception of the job of teaching, one that would make the teachers of mathematics professionals.

NONPROFESSIONAL ACTIONS OF TEACHERS

As a high school and college teacher of mathematics I never doubted that my colleagues and I were professionals. In fact, in 1970 I described a teacher as a professional "striving to improve the odds in his favor . . . [who] searches the literature, asks questions of authorities, attends professional meetings, and so on, in the hope of finding help" (Romberg, 1970, p. 56). However, during the decade of the 70s I became skeptical. Let me share with you two anecdotes to illustrate why my beliefs about the professionalism of teachers changed.

First, in the early 1970s when working with elementary schools that were adopting and helping define Individually Guided Education (IGE) (Klausmeier, Rossmiller & Saily, 1977), I was surprised at the reaction of teachers, administrators, and publishers to the motivation research conducted at the

Wisconsin Research and Development Center. Then, as now, many teachers complained about the need to motivate their students to learn, but they dismissed out of hand the well-researched motivation techniques dealing with goal-setting, tutoring, and so on (Klausmeier, Jeter, Quilling, Frayer, & Allen, 1975). I found it difficult to understand the reactions of the teachers with whom I was working. In medicine when a new treatment is developed and shown to be effective it is quickly and enthusiastically adopted. Why is that not the case in education?

Second, when studying the implementation of *Developing Mathematical Processes* (Romberg, Harvey, Moser, & Montgomery, 1974, 1975, 1976) I observed numerous teachers adapting activities that had been carefully developed. The adaptations were made for a variety of reasons. Often when interviewed, teachers justified the changes in terms of management (the class would be noisy) or perceived ability of their students (I have the low group of students). On other occasions the changes that were made drastically altered the mathematical intent of the lesson. For example, a sequence of activities on measuring length was designed to have first-grade children measure a variety of objects in their room with arbitrary units such as pencils, cards, and links. Then the teacher was to conduct a discussion on the objectives of the topic, including the iteration process used to count the number of units, the fact that small units yielded a large number for the measure while large units yielded a small number, and that there always is some error when measuring that needs to be resolved. The discussion eventually was to lead to children seeing the need for a common unit. One teacher changed the activity so that instead of measuring real objects with arbitrary units that required iterating and counting, the children measured line segments on a work sheet with a ruler marked in inches. Furthermore, the line segments were drawn so that the measures were nearly exact. The teacher justified the changes in terms of her view of mathematics: It was wrong to give children the impression that a math problem could have several right answers, or that mathematics was ever imprecise (Stephens, 1983). A lawyer who interpreted a law incorrectly would soon have few clients. Why is it assumed to be acceptable for teachers to modify and distort lessons so that the knowledge being transmitted is wrong?

These two examples illustrate teaching situations where, in my judgement, teachers were not acting in a professionally responsible manner. In the first case, research-based techniques that could improve practice were ignored; in the second, the knowledge being distributed to students was incorrect.

My skepticism during the 1970s was based on my experience as a curriculum researcher and developer whose findings and materials had been rejected or modified by teachers. Initially I blamed teachers for failing to see the merit in those ideas. I was oblivious to the fact that under existing conditions, adoption of these ideas could only make their jobs harder. Today

I understand that the problem is more systemic. What has become apparent is that both the constraints of the workplace and the unwillingness of educators to reconceptualize the job of teaching contribute to the problem.

Because it requires a small number of adults to control a large number of students, the job of teaching is viewed as managerial. Thus, many teachers approach their job with a "work place mentality." With this perspective it is not surprising that in the motivation example it was difficult for teachers to consider discussing and setting goals for each student every month. Ten minutes per month alone with each student simply could not be arranged, no matter how beneficial. The difficulty, as Sternberg (1987) has put it, is that we find it hard to consider an alternative to "the teacher is the teacher and the student is the learner" (p. 456). In fact, it appears that the only new procedures most teachers seem to be able to handle are those that can be used within the existing organizational structure and methods of teacher-student interaction.

Corcoran (1987) stated, "teachers, when reading an article or hearing new ideas, picture what they would do in their classroom." Creating such mental models of what one would do is both essential and limiting. It is essential if teachers are to use new ideas; it is limiting if their past experience in classrooms is inadequate. The teacher using the measurement activity could not envision mathematics with controversy or error and a need for discussion and resolution; she made the activity more comfortable for her but distorted the mathematics. In her view, school mathematics is different from real mathematics.

TEACHERS AS PROFESSIONALS

What does it mean to be a professional teacher? If teachers are not professionals, what would it take to make them professionals? In order to answer these and several other related questions I have chosen to structure the presentation as follows. First, the meanings associated with the common use of the term *professionalism* are considered; this part concludes with a definition consistent with the term's current use in describing an occupation. Second, the question of what might define a profession of teaching, specifically mathematics teaching, is described and contrasted with information about current practice. Finally, I summarize the activities of two projects that are addressing the professionalism of mathematics teachers.

The Meaning of "Professionalism"

One difficulty with the current discussion of teacher professionalism lies in our inability to distinguish between the behavior of individuals within an occupation agreed to be a profession, and the behavior of members of other occupational groups. The *Oxford English Dictionary* (1971) allows one to trace the notion of a profession to its religious roots as a vocation to which one professes.

Given the roots of professionalism, it is instructive to examine the set of quite particular features that characterize religious orders. Religious orders claim access to a specific body of knowledge for their members. They are restrictive in their membership and maintain control over entry to their ranks. They claim the existence of a greater value in their work, and they look to their own leadership for direction, to the point of taking oaths of obedience. In other occupations generally thought of as professions, these characteristics have been adopted, codified, and extended.

A definition of professionalism. There are four aspects of professionalism that seem to be important: First, there is the belief that those who are professional have as a result of education, training, and experience some "professed" knowledge that sets them apart from others. In fact, they both claim monopoly on certain knowledge and continually update it as needed. Second, professionals must use that knowledge when making judgements and decisions in their occupation. Third, there are some important attributes about professional occupations that make them different from other jobs. Hall (1968), from a functionalist perspective, identified five such attributes that characterize professions:

1. Professional organization as a referent: Belief in both a formal organization for governance and the importance of informal exchange with colleagues as the major source of ideas, judgement, and identity

2. Public service: A belief that the profession performs an indispensible public service

3. Self-regulation: A belief that members should be given control over their vocation, including entry requirements and judgments of performance

4. Vocation: An inner compulsion to the profession

5. Autonomy: The freedom to make professional decisions without pressure from other professions, nonprofessionals, or employing institutions.

Finally, although in Hall's view these attributes characterize professions that are largely homogeneous communities whose members share identity, values, definitions of role, and interests, they do not, as Bucher and Strauss (1961) have argued, allow for either conflicting interests within a profession or the process of change in a profession. Thus, indicators of conflict and change comprise the fourth aspect of a profession.

Teaching as a profession

Given this description of professionalism let me now outline the characteristics of teaching that would fit the definition.

Knowledge. The authors of the Carnegie report (1986) stated, "Professional work is characterized by the assumption that the job of the professional is to bring special expertise and judgement to bear on the work at hand" (p.

36). One problem that has not been adequately addressed is that we do not understand the nature of the knowledge that underlies that special expertise. In fact, it has been argued that "education lacks a well-defined body of knowledge that is applicable to the real world of teaching" (Ornstein, 1981, p. 196). It seems intuitively plausible that the more a teacher knows, the better will be both teaching performance and student outcomes. Unfortunately and rather surprisingly, most research to date has not shown that subject-matter knowledge is at all related either to teaching performance or to student achievement (cf. Druva & Anderson, 1983). However, there are considerable problems both with the type of teacher knowledge probed (cf. Shulman, 1986) and the manner of its elicitation and representation (cf. Champagne, 1986).

Recently there has been increased interest in teacher knowledge. For example, Shulman (1986) and his colleagues at Stanford, Buchmann (1983) and her colleagues at the Institute for Research on Teaching (IRT), Conroy (1987) and my staff at the Wisconsin Center for Education Research (WCER), are all examining aspects of teacher knowledge about the mathematical or scientific content that teachers are expected to teach to their students. It is premature to discuss results of these investigations at this time; however, what is clear is that the "professed knowledge" of teaching should include at least three distinctly different but related categories:

1. knowledge of the subject (mathematics) they are to teach to students and its relationship to other content (both within and outside of mathematics);

2. knowledge of pedagogy, including an understanding of how students process, store, retain, and recall information and of how teachers contract with students for instruction; knowledge of a variety of cases (examples) for each mathematical idea; knowledge of specific instructional techniques; knowledge of instructional materials; and

3. knowledge of how to manage a complex instructional setting involving a large number of students, a variety of resources, space, and an increasingly complex instructional technology.

I have elected not to review pedagogic and managerial knowledge since these topics are well covered by other participants in this conference. Instead, I will focus only on the mathematical knowledge of teachers.

In the past the typical measure of a teacher's mathematical knowledge was based on college or university degrees, number of courses or credits, or grade point average. Such indicators do not capture a teacher's understanding of the history or philosophy of the subject, or of the notion that mathematics is a growing, dynamic discipline. The teacher in the measurement example described earlier had taken several more courses in mathematics than most elementary teachers but really did not understand measurement. In a similar manner, Lovitt and Clarke (1986) have described secondary teachers who, when teaching a clever unit on scoring olympic diving, only see it as an exercise in finding averages and not one in developing an ordinal scale for qualitative events. The epistemological question— What does it mean to know mathematics?—is important for both teachers and students.

In order to answer the epistomological question, we must first decide what mathematics we want students to know. When that is satisfactorily determined, we must consider how such knowledge is acquired, what knowledge teachers must have, and finally, the pedagogy and management appropriate for that knowledge. A brief review of the literature yielded no studies that addressed these epistemological issues with teachers.

Finally, it is assumed that professionals read professional journals and research and continue to update themselves. In one study, George and Ray (1979) found that fewer than 30% of teachers ever read or consulted a professional journal.

Use of knowledge. "Teachers are considered as professionals because . . . the demands of the job require judgement and decision making" (Romberg, 1985, p. 2). The assumption is that the knowledge teachers have is necessary for them to make informed judgements and decisions. Is that really the case? The National Advisory Committee on Mathematical Education (NACOME) (1975) commissioned a study of elementary school mathematics instruction. The picture drawn from that survey is as follows: "The median classroom is self-contained. The mathematics period is about 43 minutes long, and about half of the time is spent on written work. A single text is used in whole-class instruction. The text is followed fairly closely, but students are likely to read at most one or two pages out of five pages of textual materials other than problems. For students, the text is primarily a source of problem lists" (p. 77). Within this context, other studies commissioned by the National Science Foundation (NSF) have shown that the daily sequence of activities involved in teaching mathematics was as follows:

> First, answers were given for the previous day's assignment. A brief explanation, sometimes none at all, was given of the new material, and problems were assigned for the next day. The remainder of the class was devoted to students working independently on the homework while the teacher moved about the room answering questions. The most noticeable thing about math classes was the repetition of this routine. (Welch, 1978, p. 6)

Clarke (1984) has given the following description of mathematics class from one student.

> Wait outside if she's not there. Come in if she's there. Sit down. And she tells us what we're gonna do. And she'll probably write up a few examples and notes on the board. Then we'll either get sheets handed out or she'll write up questions on the board. Not very often. We mainly get a textbook. We'll get pages. She'll write up what work to do, page number and exercise. And if you finish quick you may get an activity sheet. And that's about what happens. (p. 38)

From these complementary pictures of the typical mathematical classroom and the job teachers actually perform, it is hard to argue that the job of teaching requires professional judgement. The teacher's job is not related to a conception of mathematical knowledge to be taught, to an understanding of how learning occurs, or to knowledge of the likely outcomes of various instructional actions. Special knowledge about mathematics or findings from recent research on learning or teaching have little relevance because the judgments and decisions teachers make do not depend on such knowledge. Furthermore, even if teachers had the appropriate "professed" knowledge to make reasonable judgements and decisions, they could not, for "teachers work in an environment suffused with bureaucracy. Rules made by others govern their behavior at every turn" (Carnegie, 1986, p. 39). Finally, it is assumed that professionals change their practice based on new knowledge (see the first example). In fact, it is in this area that teachers have been most disenfranchised. Teachers say they support and see the need for change but are left out of decisions (National Governors' Association Center for Policy Research and Analysis, 1986).

Attributes of professions

In this section a brief review of the limited research associated with teaching as a profession is examined. Sociologists interested in classifying occupations have long been interested in the attributes that distinguish professions from another occupations. In order to study professions, Hall's (1969) model, described earlier, will be used. This model is composed of both structural and attitudinal characteristics (professional organization as a referent, public service, self-regulation, vocation, and autonomy). Hall developed a questionnaire to gather information from members of several professional groups; he sought to describe the common features and variability across professions, and to study the relationship between professionalism and bureaucratization. Several other researchers used Hall's instrument to study teachers as professionals (Weil & Weil, 1971; Alutto & Belasco, 1973; Ornstein, 1976, 1981). Snizek (1972) revised Hall's instrument, which then was used by Lam (1982, 1983) in several studies of teachers (1982, 1983). In 1986, Donovan and Romberg revised Snizek's instrument for use with mathematics teachers in the Urban Mathematics Collaborative Project (Romberg, Webb, Pittelman, & Pitman, 1987). For that questionnaire, "collegiality" replaced "professional organization as a referent" to encompass within-school and between-school relationships as well as those with a professional organization.

Collegiality refers to relationships with others. "The improvement and professionalization of teaching depends ultimately on providing teachers with opportunities to contribute to the development of knowledge in their profession, to form collegial relationships beyond their immediate working environment, and to grow intellectually as they mature professionally" (Holmes Group, 1986, p. 66). The importance of professional collegiality has been well documented.

In a summary of research, Little (1982) found that a high level of collegiality was related to a positive school environment which in turn was the best predictor of school effectiveness. This collegiality involves in-school, across-school, and outside-of-school relationships. Little also identified four critical practices that characterize within-school collegiality:

1. teachers engage in frequent, continuous, and increasingly precise and concrete talk about teaching;

2. teachers are frequently observed by other teachers and provided with useful critiques of their teaching;

3. teachers plan, design, research, evaluate, and prepare teaching materials together; and

4. teachers teach each other the practice of teaching.

In an examination of the isolation of teachers, Copeland and Jamgochian (1985) found that networks of professional colleagues provide significant pathways for communication and for the adoption of innovations.

When examining the subgroup nature of collegial relationships among teachers, Ginsburg and Pearson (1981) found that the social relationships among teachers are an important part of being a teacher. It is the teacher's colleagues who control his or her induction into the profession, the conception of what it means to be a teacher, and what is valued.

Hall (1969) found that one important source for outside-of-the-school relationships is professional organizations. In most other occupations that claim to be professional, the professional organizations, in large part, dictate admission into the profession, expulsion from the profession, and the standards of the profession. From a sample of over 1200 teachers, Lam (1982) found that there was a very high agreement on the importance of professional organizations.

Bucher and Strauss (1961) saw colleagueship as the most sensitive indicator of change in professions. They argued that professionals belong to "circles of colleagueship" (p. 330) that provide selective influence to members. Change in professions is apparent when circles of colleagueship change.

These are only a few of the studies on this topic, but they provide sufficient evidence that collegiality is very important. However, it is also clear that there are several impediments to the development of such collegiality in most schools. Copeland and Jamgochian (1985) found that the primary reasons teachers give for the lack of collegial relationships are time and schedule constraints, and administrators' concepts of the nature of the teacher's job. Grant and Sleeter (1985) found that the "middle class conservative" view of teaching held by most teachers from behind their desks limits their vision. Many teachers consider current practice as basically good and see no need to change the status quo.

Time seems to be a major problem. The Carnegie group described the problem well.

> Fundamental to our conception of a workable professional environment that fosters learning is more time for all professional teachers to reflect, plan and discuss teaching innovations and problems with their colleagues. Providing this additional time requires additional staff to support the professional teachers, technology that relieves teachers of much routine instructional and administrative work, a radical reorganization of work roles to make the most efficient use of staff in a collegial environment, and a new approach to the use of space. (1986, p. 60)

Recently, several authors (e.g., McDonald & Naso, 1986; Sternberg, 1987) have argued that we need to consider teachers as learners. They need to have and be collegial advisors. In particular, this collegial advisor role is being proposed as an appropriate relationship between teachers and students, particularly with respect to such issues as the use of technology and the development of higher-order thinking skills.

In summary, in spite of the overwhelming evidence of the importance of collegial relationships, teachers are, for the most part, isolated from each other in the same buildings, from teachers in other buildings or districts, and from other mathematicians. One way to foster change would be to create new collegial opportunities.

Public service implies that "the occupation is indispensable and that it benefits both the public and the practitioner" (Hall, 1969, p. 82). Teaching obviously fits this definition. Hall (1969) found that teachers ranked higher on public service than any of 27 other occupational groups. Lam (1982) found that all teachers felt public service was an important characteristic of their job. Pride in service to the community seems to be common among all teachers.

Self-regulation "involves the belief that, since the persons best qualified to judge the work of the professional are his fellows, colleague control is both desirable and practical" (Hall, 1969). On this property of professions, the evidence about its importance to teaching is ambiguous. Both Hall (1969) and Lam (1982) found that teachers did not rate this characteristic as important. On the other hand, Duke, Showers, and Imber (1980) found that most teachers felt self-regulation was too hard. Teachers' perceptions of colleague control were related to the costs and benefits of involvement in school decision making; the teachers they interviewed rated the potential costs as low and the potential benefits as high. However, the teachers were hesitant to become involved because they saw little possibility that their involvement would actually make a difference. Bureaucratic structures did not allow them to follow up on their involvement.

Vocation refers to "the dedication of the professional to his work" (Hall, 1969, p. 82). Although it is hard to conceive of teaching without a sense of vocation, what little evidence exists shows that most teachers do not agree with its importance. Again, both Hall (1969) and Lam (1982) found this to be the case. Obviously, the aphorism that "those who can, do; those who can't, teach," evinces a high level of skepticism about the avocation for many teachers.

Autonomy involves the belief that "the practitioner ought to be allowed to make his own decisions without external pressure from clients, from others who are not members of his profession, or from his employing organization" (Hall, 1969, p. 82). Teachers have, in one sense, a high degree of autonomy inside their own classrooms, but as a group they have been largely disenfranchised with respect to many decisions about issues such as curricular goals, methods of assessment, and scheduling. Furthermore, given current legislative leanings, they are in danger of losing whatever professional autonomy they once enjoyed. It is also very clear that teachers strongly believe in the importance of this professional characteristic (Hall, 1969; Lam, 1982).

However, Kreis and Brockopp (1986), when reviewing teacher job satisfaction, found that teachers' autonomy is limited and contradictory. Teachers have authority over students, but often lack authority over their school-wide environments and over themselves. For example, in contrasting students preparing to teach with veteran teachers, Vavrus (1978) found that the students assumed they would have control over their environments, while the veterans said they had no such control. Kreis and Brockopp (1986) found that teachers expect autonomy only in their classrooms.

In summary, it seems apparent that, while there is currently considerable interest in improving the professional status of teachers—in particular mathematics teachers (e.g., Carnegie, 1986; Holmes Group, 1986)—teachers are not now professionals. Furthermore, while suggestions are being made about activities that would make professionalism possible, the suggestions are not based on a well-grounded theory of professional practice.

Examples of Programs Designed to Influence the Professional Status of Mathematics Teachers

In this section I discuss two programs I have been directly involved with during the past two years: the Urban Mathematics Collaborative Project funded by the Ford Foundation (Romberg, Webb, Pittelman, & Pitman, 1987) and the Mathematics Curriculum and Teaching Project funded by Australia's Curriculum Development Centre (Lovitt & Clarke, 1986).1 Both projects deal directly with teachers and provide them with new knowledge and nondirective support to utilize that knowledge. However, the projects have addressed different aspects of knowledge and used different strategies.

The *Urban Mathematics Collaboratives* are located in eleven urban centers across the U.S. (Cleveland, OH; Durham, NC; Philadelphia, PA; Pittsburgh, PA; Los Angeles, CA; Memphis, TN; New Orleans, LA; San Diego, CA; San Francisco, CA; St. Louis, MO; and Minneapolis-St. Paul, MN). In addition, there is a documentation project to monitor the activities of the collaboratives, and a technical assistance project. The intention of the Ford Foundation was to provide a framework for enhanced teacher professional activities. The initial objective of the Foundation was to provide nondirective support that would enable each collaborative to develop an organizational framework and to choose its own focus of interest. As the effort continues, collaboratives are supposed to focus on the effects of the developing networks on the professional lives of the participating teachers and on the identification of issue-based outcomes. The Foundation's intention in this effort is consonant with the recommendations of the Conference Board of the Mathematical Sciences (1984):

> The Conference recommends the establishment of a nationwide collection of local teacher support networks to link teachers with their colleagues at every level, and to provide ready access to information about all aspects of school mathematics. (p. 5)

The broad sense in which the term *colleague* is used is exemplified by the objectives "strongly endorsed by the conference":

- To extend the sense of professionalism among teachers by building a support system that links them to colleagues in the mathematical sciences, inside and outside of the schools;

- to provide teachers at all levels with colleagues upon whom they can call for information concerning any aspect of school mathematics; and

- to enable teachers to enlarge their views of mathematics, their sources of examples, and their repertoire of classroom skills in communicating mathematics. (CBMS, 1984, p. 15)

It has been the aim of the Ford Foundation to assist in the establishment of networks in which mathematics teachers can participate as colleagues with mathematicians in business, government, higher education, and industry. In the collaboratives, mathematics teachers are expected to be participants rather than clients in the current reform effort.

We have three preliminary findings based on our observations. First, the highest priority for each collaborative during the first two years has been to intellectually challenge a group of secondary mathematics teachers. As the job of teaching tends to become repetitive, many of the teachers had not been intellectually challenged since they had been in college many years before. The activities offered by the various collaboratives, such as the opportunity to solve mathematical problems, to work on a computer, to assist in an industry, and to explore physics problems at the Exploratorium, were very exciting, and greatly appreciated by the teachers. Equally important as the activities themselves was the opportunity for teachers to discuss their plans and follow-up activities with their colleagues. This intellectual refreshment has been the most important outcome of every collaborative.

Second, an important feature of the activities in each collaborative has been that the teachers are being treated differently. Teachers were not given already developed materials or procedures that they were expected to implement. Instead, the expectation was that they would become more familiar with issues and with the problems associated with those issues. Teachers have been viewed as partners (concerned professionals) in the educational process, and their ideas are respected. Many teachers initially viewed this approach suspiciously, since in the past they had been treated as passive clients rather than as colleagues. The corporate dinners and the visits to industries and universities, however, gave teachers an opportunity to interact with industry and university mathematical scientists. Teachers in all the collaborative sites appreciated both the opportunity to be with industry and university mathematicians and the respect they were shown.

Third, teachers in each site are beginning to become aware of the social and political problems related to curriculum and instructional reform. They recognize the need for time to reflect, the need for additional training, and the problems inherent in developing materials for implementation in a classroom. Traditionally, teachers have been treated as conduits in a system, with the assumption that the only responsibility of the teacher was to teach students in a classroom. However, if teachers are to become real partners in a reform effort which would involve developing materials, trying them out, and discussing ideas with others, they will need time to plan, to develop, to reflect, and to evaluate. Furthermore, many teachers are now beginning to realize that their mathematics backgrounds are outdated, with their training often reflecting only the content of the NSF Institute programs of the 1960s. After seeking new topics from industry, teachers are asking for good examples they can use in their classrooms. All teachers realize that instructional materials (texts, software, tests, etc.) will need to be developed, although not every teacher or school district needs to develop these materials independently.

The *Mathematics Curriculum and Teaching Project* (MCTP) has been jointly funded by the Australian federal government, through its Curriculum Development Centre, and all of the Australian states. It is an attempt to get teachers to help teachers. MCTP saw ongoing support as the biggest single need of teachers as they seek to address, in classroom practice, the concerns expressed in the current efforts to reform mathematics teaching. The need for support was stated by Lovitt and Clarke (1986).

> One outstanding observation of the emergence of new and important growth areas in mathematics education is the need to involve and inform teachers about the consequences and characteristics of their classroom implementation. Teachers should be knowing and active partners in the process of analyzing the suitability of new ideas for inclusion into the maths curriculum. Not to do so is to sow the

seeds of failure of implementation or dissemination strategies. This was most notably seen in the introduction of the "new maths" during the late sixties, and arguably true of many current attempts to implement "problem solving." With the current range of growth areas such as the use of language, problem solving, technology and applications, teachers can rightfully say, "We don't have a clear strong image of what these new emphases would look like in classroom practice— and we don't have the confidence, experience or resources to put them into practice." (p. 2)

The focus of MCTP has been to engage teachers in the debate about the values of new or emerging concerns by supporting them in actively exploring these issues in their own classrooms and in association with their own colleagues. In this way, the program can help teachers to deepen their understandings of the way students learn and, in response, to develop appropriate teaching strategies.

MCTP has supported the teachers in two closely related ways:

1. by provision of excellent, well-documented *images of classroom practice* within these new areas of concern. These images act as illustrations or templates and represent the most promising activities collected from across Australia.

2. by provision of excellent, well-documented procedures found to have been successful at supporting teachers in such processes. These professional development procedures will likewise be the best of current practice from across the country.

In simple terms, this means that in teachers' chosen interest areas, MCTP is seeking to provide the best illustrations available, to encourage teachers to trial these in their own classrooms, and then to provide opportunity for reflection, so that in the light of experience, any underlying principles can be recognized and adopted into the teachers' own personal "repertoires." (Lovitt & Clark, 1986, pp. 2-3)

As an external reviewer (a "friendly critic"), I was delighted to see good teachers helping other teachers. All of the materials were developed by teachers; teachers illustrate practices, coach others, and so forth. It is a model of teacher change that we need to consider.

In summary, these two projects illustrate the kind of efforts that are needed. Given the clamor for reform in the teaching of mathematics, there are undoubtedly many other efforts that are focused on making teachers more professional. It is clear that any reform in mathematics teaching and education that is to take place in the 1990s must actively involve teachers in the process of reform in order for it to be effective. Reform efforts that view the teacher as a passive recipient, such as attempts to develop "teacher-proof" curricula, are likely to fail. In contrast, reforms that take into account teachers' beliefs, perspectives, and knowledge; that actively involve teachers in planning and decision making; and that treat teachers as professionals are more likely to succeed. Instructional design should not be prescriptive;

instead, it should provide a framework, a knowledge base, alternative materials for instruction, and a variety of appropriate assessment techniques that enable teachers to make informed decisions.

A VISION ABOUT PROFESSIONAL TEACHERS

For mathematics teachers to be true professionals, two things are needed. First, a new conception of mathematical literacy needs to be adopted. Second, "school systems based on bureaucratic authority must be replaced by schools in which authority is grounded in the professional competence of the teacher, and where teachers work together as colleagues, constantly striving to improve their performance" (Carnegie, 1986, p. 55).

Mathematical Literacy

The Carnegie Forum (1986) echoed an international position on the task of teaching when it stressed that students must be active learners, busily engaged in the process of bringing new knowledge and new ways of knowing to bear on a widening range of increasingly difficult problems. The focus of schooling must shift from the passive acquisition of facts and routines to the active application of ideas to problems. Minimal literacy acquired early in life through passive methods is no longer enough. The personal, national, and global problems to be faced and solved require adults who are able to continue to learn and adapt because they have confidence in their own personally-created and firmly-founded understanding (cf. Committee on Inquiry into the Teaching of Mathematics in Schools, 1982). The situation requires that the industrial notion of literacy as conditioning for predictable reaction be revised to the notion of literacy as the capability for innovative adaptation.

This need for literacy to support continued learning draws attention to Resnick and Resneck's (1977) distinction between high and low literacy. Under the industrial model, teaching the low-literacy majority was essentially a matter of acculturing and training children to fit a system that required punctuality, obedience, predictable performance, and minimal literacy. There are obvious connections between this model, stimulus-response theory, and teaching and evaluating according to behavioral objectives. At best, the industrial model amounted to Skemp's (1983) notion of instrumental learning—the acquisition of a repertoire of fixed plans.

By contrast, present demands require a high-literacy model for all students. For example, the Conference Board of the Mathematical Sciences (1984) recommended that "no student except the severely handicapped should be able to choose courses in a way which, before grade 11[,] makes it impossible for him/her to move to a college preparatory curriculum" (p. 11). A clear implication is that all students should enter the eleventh grade with a high-literacy preparation in mathematics capable of supporting continuing education. This statement clearly rejects the notion of literacy as minimal training and replaces it with the ability to continue learning.

"People only begin to learn when they go beyond what they are taught and teach themselves, reflectively taking initiatives" (Greene, 1982, p. 32). Thus, high literacy is constructive, not passive, because it necessitates relational understanding, which entails "building up a conceptual structure . . . from which its possessor . . . can produce an unlimited number of plans" (Skemp, 1979). It is a dynamic concept of literacy as supporting quest, accommodation to new environments or changed conditions, and the active creation of knowledge. By contrast with the passivity and domination implicit in low literacy, high literacy has an empowering quality (Greene, 1982) and political implications (Hirsch, 1983). High literacy is implicit in the repeated demand for adults who can think, learn, and solve problems.

However, the fundamental problem is that the existing educational system is a tightly coherent ecosystem (Futrell, 1986), which is highly efficient at the function for which it was intended: mass low literacy. A change to mass high literacy, supporting adults who can learn, adapt, innovate, and create knowledge, is unlikely unless the entire structure is reconceptualized around that purpose. Thus, the crux of reform is how to establish an equally coherent ecosystem for high literacy, in particular, one that includes notions of teaching for high literacy.

For teachers, this means several things. First, they must learn to encourage students to reflect on and argue their intuitions. Such activity is essential to the process of mathematizing and to the development of mathematical attitudes. Encouraging students to explain how they "know" and to persuade and convince others of that knowledge, is an important ingredient of mathematical expertise. It requires written logs of accumulating information that supports a belief, reflection on and discussion of that record, and a more formal report to present beliefs and grounds (cf. Biggs, 1985).

Contemplating extended student dialogue, Lochhead (1979) made the following points:

1. Teachers talk too much . . . [they] "must learn to shut up and listen." One of the best ways to do this is to encourage serious dialogue between students.

2. What at first appears to be poor performance may on closer examination turn out to be impressive work. . . .

3. During learning, performance may temporarily decrease; only over the long term is there a steady increase. . . .

4. Students can construct mathematical concepts and formulas if given time to do so. Most current instruction systematically denies students this opportunity. . . .

5. The correct use of terminology is a natural consequence of understanding but does not result from skill in the correct use of terminology. . . .

6. Promoting constructive dialogue is not easy. It requires a thorough understanding of the student's stage of development and a deep understanding of the material to be taught. (p. 177)

Second, teachers must allow students to explore situations embodying a variety of mathematical questions, each of which may be addressed by different methods and described in different ways. This places rigorous demands on teachers' abilities to adapt instruction to the requirements of the children. They must not only have an intensive understanding of their subject "but must understand in a pedagogically reflective way; they must not only know their own way around a discipline, but must know the 'conceptual barriers' likely to hinder others" (Hawkins, 1974, p. 8).

Third, if students are to be free to identify and pursue their own questions, it is important for teachers to be able to diagnose the condition of a student's cognitive structure in order to help identify questions that will be challenging but not defeating.

Since learning requires making sense of things and creating meaning, and establishing relationships between new information and what is already known, teachers must understand the difficulties faced by students. Difficulties depend upon failure to relate the new to what is already known, deficiencies in knowledge, and incorrect elements in the knowledge possessed (Simpson, 1985). Bell (1982) described diagnostic teaching as emerging from research on understanding. Even if a student scores an acceptable percentage of correct answers on a test aimed at a spectrum of understanding in mathematics, the score offers a misleading picture of that student's understanding. It may well conceal serious misconceptions that are carried on and become obstacles to subsequent learning. Therefore,

> we need to take the diagnosis and treatment of pupils' mistakes far more seriously than we do—in fact, in general we need to do less direct teaching and concentrate on finding out what they think in relation to the problems on hand, discussing their misconceptions sensitively, and giving them situations to go on thinking about which will enable them to readjust their ideas. (Bell, 1982, p. 7)

Consequently, teaching should be directed less toward assessment than toward a strategy for teaching that contributes to development of childrens' understanding. By definition, it requires starting from the structure of the child's knowledge. The information on which to base diagnosis may be acquired through observation, questioning, and conversation. Conversation with the student is crucial to diagnosis, as is reflection on the products of student activity. A key strategy is to prompt children to develop their knowledge structures by deliberately provoking predictive inadequacy. Perturbation of existing structures is accomplished through critical tasks that embody known misconceptions.

Finally, the oral tradition of teaching-as-telling created a clearcut distinction between learning and teaching and, therefore, between the corresponding responsibilities and authorities of teachers and students. Integrating theories of teaching and learning would suggest that, if children are to take an active involvement in their own learning, these social duties and relationships also need to be integrated.

In summary, the teaching of mathematics has an inescapable social context that requires the redefinition of mass literacy from basic training to the universal empowerment of human potential. Given the goal of long-term understanding in support of life-long learning, a radically different, but strongly supported, picture of school mathematics teaching is needed. Succinctly, teaching for long-term learning and the development of knowledge structures requires teachers to create epistemic situations in which children can explore problems, create structures, generate questions, and reflect on patterns. It requires teachers with the academic and pedagogical knowledge to provide for flexible approaches, encouraging informal and multiple representation while fostering the gradual growth of mathematical language. It requires:

1. teachers who can diagnose difficulties and devise questions to promote progress through cognitive conflict;

2. teachers who can maintain a collaborative atmosphere leading to increasing independence on the part of students while also doing everything else;

3. careful reflection on strategies for cognitive motivation;

4. recognition of the epistemic, cognitive (cf., Skemp, 1979), and social commonalities of teaching and learning.

New School Organizations

In order to improve schooling, "the public must offer teachers a professional work environment. That means a better salary, but more to the point, it means more resources to do the job. It means giving teachers a real voice in educational decisions . . . and more control over their time" (National Governors' Association Center for Policy Research and Analysis, 1986, p. 36). Schools as they currently exist must change both in physical and social structure if this kind of environment is to be created. I am not sure what the physical structure of schools should become, perhaps the Minnesota Plan is a reasonable starting point (cf. Berman, 1985), but I have some suggestions about the social structure.

1. Make teaching a full-time job. The United States is the only major country that hires teachers for 9 to 10 months.

2. Reduce class load. Three hours of instruction per day is plenty, since teachers need time to reflect, to plan, and to work with individual students.

3. Organize differentiated scheduling and staffing for mathematics classes. Master teachers need not meet every class every day for every activity. Mathematics specialists are needed for all classes, but may only engage in direct instruction 2 or 3 days a week; other adults must be responsible for other instructional activities. Also, there need to be clear career ladders so that teachers see a professional future.

4. Consider alternate means of relating teacher accountability and autonomy. New incentives and opportunities for teachers in exercising autonomy in schools need to be developed. But autonomy does not mean license; coupled with such freedom comes responsibility and inevitably accountability. For example, the "Private-Practice Teaching in Public Schools" program (Olson, 1986) now being tried out in Minnesota is one approach that needs to be considered.

5. Create and foster collegial relationships between teachers and others interested in the teaching and learning of mathematics.

Other Recommendations

Given the perspective I have taken in this paper about mathematical literacy and school organization there are other recommendations which need to be made with respect to the preparation and development of professional mathematics teachers.

1. Ensure excellent preservice and in-service education, congruent in style with the quality of teaching expected.

2. Provide for teachers to constantly expand their own knowledge of the domain through such things as sabbaticals, summer scholarships for foreign study, in-service programs, computer conferencing with experts, and so on, placing no restrictions on the directions of investigation.

3. Provide for constant electronic collegiality.

4. Develop a framework for rigorous self-regulation.

CONCLUSION

The question posed in the title of this paper, "Can teachers be professionals?" can now be answered: Emphatically, yes! However, to do so will take time and considerable resources. In addition, the concerted effort and commitment of a large number of persons will be needed to make it a reality in the coming decade.

NOTES

[1]In the original paper prepared for this conference I also discussed the Cognitively-Guided Instruction Project (Carpenter, Fennema, & Peterson; 1985). It has been omitted here because it is discussed by Peterson in this volume.

REFERENCES

Alutto, J. A., & Belasco, J. A. (1973). Patterns of teacher participation in school system decision making. *Educational Administration Quarterly, 9*(1), 27-41.

Bell, A. W. (1982). Diagnosing students misconceptions. *The Australian Mathematics Teacher, 38*(1), 6-10.

Berman, P. (1985). The next step: The Minnesota plan. *Phi Delta* Kappan, 67, 188-193.

Biggs, E. (1985). *Teaching mathematics 7-13: Slow learning and able pupils*. Windsor, UK: NFER-Nelson.

Bucher, R., & Strauss, A. (1961). Professions in process. *American Journal of Sociology, 66*, 325-334.

Buchmann, M. (1983). *The priority of knowledge and understanding in teaching* (Occasional Paper No. 61). East Lansing, MI: Institute for Research on Teaching, Michigan State University. (ERIC Document Reproduction Services No. ED 237 503)

Carnegie Forum on Education and the Economy. (1986). *A nation prepared: Teachers for the 21st century. The Report of the Task Force on Teaching as a Profession.* New York: Author.

Carpenter, T. P., Fennema, E., & Peterson, P. (1985, March). *Cognitively guided instruction: The application of cognitive and instructional science to mathematics curriculum development.* Paper presented at the International Conference on Mathematics Education, University of Chicago.

Champagne, A. B. (1986, April). *Science teacher quality: A cognitive perspective.* Paper presented at the annual meeting of the American Educational Research Association, San Francisco.

Committee on Inquiry into the Teaching of Mathematics in Schools. (1982). *Mathematics counts* (The Cockroft Report). London: Her Majesty's Stationery Office.

Clarke, D. (1984). Secondary mathematics teaching: Towards a critical appraisal of current practice. *Vinculum* (Magazine of the Mathematical Association of Victoria, Australia), *21(4)*, 22-42.

Conference Board of the Mathematical Sciences. (1984). *New goals for mathematical sciences education.* Washington, DC: Author.

Conroy, J. (1987). Monitoring school mathematics: Teachers' perceptions of mathematics. In T. A. Romberg & D. M. Stewart (Eds.). *The monitoring of school mathematics: Background papers. Volume 3: Schooling, teachers and teaching.* (Program Report 87-3, pp. 75-89). Madison, WI: Wisconsin Center for Education Research.

Copeland, W. D., & Jamgochian, R. (1985). Colleague training and peer review. *Journal of Teacher Education, 36*(2), 18-21.

Corcoran, C. (1987, January). Comment made at NCTM Research Agenda Meeting, San Diego, CA.

Donovan, B., & Romberg, T. A. (1986). *UMCP: Teacher Survey II.* Madison, WI: Wisconsin Center for Education Research.

Druva, C. A., & Anderson, R. D. (1983). Science teacher characteristics by teacher behaviour and by student outcome: A meta-analysis of research. *Journal of Research in Science Teaching, 20*, 467-479.

Duke, D., Showers, B., & Imber, M. (1980). Teachers and shared decision making: The costs and benefits of involvement. *Educational Administration Quarterly,* 16(1), 93-106.

Futrell, M. H. (1986). Restructuring teaching: A call for research. Educational Researcher, 15(10), 5-8.

George, T. W., & Ray, S. (1979). Professional reading—Neglected resource—Why? *The Elementary School Journal, 80*(1), 29-33.

Ginsburg, M. B., & Pearson, J. P. (1981, January). *The structure of colleague relations in an urban elementary school.* Paper presented at the Southwest Educational Research Association, San Antonio, Texas.

Grant, C. A., & Sleeter, C. E. (1985). Who determines teacher work: The teacher, the organization, or both? *Teaching and Teacher Education, 1*, 209-220.

Greene, M. (1982, January). Literacy for what? *Phi Delta Kappan, 63*, 326-329.

Hall, R. H. (1968). Professionalization and bureaucratization. *American Sociological Review, 33*(1), 92-104.

Hall, R. H. (1969). *Occupations and the social structure.* Englewood Cliffs, NJ: Prentice Hall.

Hawkins, D. (1974). *The Informed Vision: Essays on Learning and Human Nature.* New York: Agathon.

Hirsch, E. D. (1983). Cultural literacy. In A. M. Lesgold & F. Reif (Eds.), *Computers in education: Realizing the potential* (pp. 206- 215). Washington, DC: U.S. Department of Education.

Holmes Group. (1986). *Tomorrow's teachers: A report of the Holmes Group.* East Lansing, MI: Author.

Klausmeier, H. J., Jeter, J. T., Quilling, M. R., Frayer, D. A., & Allen, P. S. (1975). *Individually guided motivation*. Madison, WI: Wisconsin Research and Development Center for Cognitive Learning.

Klausmeier, H. J., Rossmiller, R. A., & Saily, M. (Eds.). (1977). *Individually guided elementary education: Concepts and practices*. New York: Academic Press.

Kreis, K., & Brockopp, D. Y. (1986). Autonomy: A component of teacher job satisfaction. *Education, 107*(1), 110-115.

Lam, Y. L. J. (1982). Teacher professional profile: A personal and contextual analysis. *The Alberta Journal of Educational Research, 28*(2), 122-134.

Lam, Y. L. J. (1983). Determinants of teacher professionalism. *The Alberta Journal of Educational Research, 29*(3), 168-179.

Little, J. W. (1982). Norms of collegiality and experimentation: Workplace conditions of school success. *American Educational Research Journal, 19*, 325-340.

Lochhead, J. (1979). On learning to balance perceptions by conceptions: A dialogue between two science students. In J. Lochhead & J. Clement (Eds.). *Cognitive process instruction: Research on teaching thinking skills* (pp. 147-178). Philadelphia: Franklin Institute Press.

Lovitt, C., & Clarke, D. (1986). *First year report*. Canberra, Australia: Mathematics Curriculum and Teaching Program.

McDonald, J. P., & Naso, P. (1986). *Teacher as learner: The impact of technology*. Cambridge, MA: Educational Technology Center, Harvard Graduate School of Education.

National Advisory Committee on Mathematical Education. (1975). *Overview and analysis of school mathematics, K-12*. Washington, DC: Conference Board of the Mathematical Sciences.

National Governors' Association Center for Policy Research and Analysis. (1986, August). *Time for results: The governors' 1991 report on education*. Washington, DC: Author.

Olson, R. A. (1986). *Private-practice teaching in public schools*. Minneapolis, MN: Private Practice Advisors.

Ornstein, A. C. (1976). Characteristics of the teaching profession. *Illinois Schools Journal, 56*(4), 12-21.

Ornstein, A. C. (1981). The trend toward increased professionalism for teachers. *Phi Delta Kappan, 63*, 196-198.

Oxford English Dictionary, (1971). Oxford, England: Oxford University Press.

Resnick, D. P., & Resneck, L. B. (1977). The nature of literacy: An historical exploration. *Harvard Educational Review, 47*, 370-385.

Romberg, T. A. (1970). Curriculum, Development, and Research. In M. F. Rosskopf (Ed.), *The teaching of secondary school mathematics* (pp. 56-88). Reston, VA: National Council of Teachers of Mathematics.

Romberg, T. A. (1985). Research and the job of teaching. In T. A. Romberg (Ed.), *Using research in the professional life of mathematics teachers* (p. 2-7). Madison, WI: Wisconsin Center for Education Research.

Romberg, T. A., Harvey, J. G., Moser, J. M., & Montgomery, M. E. (1974, 1975, 1976). *Developing mathematical processes*. Chicago: Rand McNally.

Romberg, T. A., & Price, G. G. (1983). Curriculum implementation and staff development as curriculum change. In G. A. Griffin (Ed.), *Staff development: Eighty-second yearbook of the National Society for Study in Education* (Part 2, pp. 154-184). Chicago: University of Chicago Press.

Romberg, T. A., Webb, N., Pittelman, S., & Pitman, A. (1987). *1986 annual report to the Ford Foundation: The urban mathematics collaborative project* (Program Report 87-4). Madison, WI: Wisconsin Center for Education Research.

Shulman, L. S. (1986). Those who understand: Knowledge growth in teaching. *Educational Researcher, 15*(2), 4-14.

Simpson, M. (1985). Diagnostic assessment and its contribution to pupil's learning. In S. Brown & P. Munn (Eds.). *The changing face of education 14 to 16: Curriculum and assessment* (pp. 69-81). Philadelphia: NFER-Nelson.

Skemp, R. R. (1979). *Intelligence, learning and action: A foundation for theory and practice in education*. New York: Wiley.

Skemp, R. R. (1983). The school as a learning environment for teachers. In J. Bergeron & N. Herscovics (Eds.), *Proceedings of the fifth annual meeting of the North American Chapter of the International Group for the Psychology of Mathematics Education* (pp. 215-222). Montreal, Canada.

Snizek, W. E. (1972). Hall's professionalism scale: An empirical reassessment. *American Sociological Review*, 37(1), 109-114.

Stephens, W. M. (1983). *Mathematical knowledge and school work: A case study of the teaching of Developing Mathematical Processes*. Madison, WI: Wisconsin Center for Education Research.

Sternberg, R. J. (1987, February). Teaching critical thinking: Eight easy ways to fail before you begin. *Phi Delta Kappan*, 68, 456- 459.

Vavrus, M. J. (1978). *The relationship of teacher alienation to school workplace characteristics and career stages of teaching*. Unpublished doctoral dissertation, Michigan State University, East Lansing, MI.

Weil, P. E., & Weil, M. (1971). Professionalism: A study of attitudes and values. *Journal of Teacher Education*, 22, 314-318.

Welch, W. (1978). Science education in Urbanville: A case study. In R. Stake & J. Easley (Eds.), *Case studies in science education*. Urbana, IL: University of Illinois.

Pervasive Themes and Some Departure Points for Research into Effective Mathematics Teaching

Marilyn Nickson
Essex Institute of Higher Education

The suggestion that "all identifications take place within horizons that imply a specific social world" (Berger & Luckman, 1966, p. 152) offers an apt description of three levels of consideration in this volume. First, the focal point for the conference represented in these papers resulted from a need to consider research perspectives within the horizons of mathematics education at a national level. However, at a second level this specific social world of research in mathematics education becomes, within itself, more specific and contains areas bound by seemingly different horizons as is clearly evident from the variety of concerns in the studies presented here. Finally, there is identification at the personal level where one selects particular aspects from this spectrum of content for consideration, a process which necessarily reflects the horizons of the individual and a yet more specific social perspective. What follows, therefore, reflects such a selection that is limited not just by restrictions of space but also, inevitably, by individual perceptions of what was of particular concern at this conference which encompassed a wide range of interests within the world of research in mathematics education.

Teachers of Mathematics

Mathematics teachers and various aspects of their role have provided a strong focal point for the studies and discussion under consideration here, which logically reflects the theme of *effective mathematics teaching* being addressed. However,it is not only the act of teaching that has been considered; wider issues relating to the various roles teachers adopt and factors contributing to their individuality and professionalism are also raised. References have been made to teachers as decision makers, collaborators, experts and makers-of-knowledge either in papers or in discussion. References have also been made to teachers' isolation, to their autonomy, and to their beliefs. This range of concerns suggests a spectrum at one end of which interest lies explicitly with the act of teaching and at the other, relates more to teachers as individuals within the professional context in which they work. For example, both Leinhardt and Berliner have studied the expert teacher, while Romberg has explored the position of teachers as professionals and Bauersfeld has considered theoretical perspectives of classroom interaction. Perhaps the most fundamental of the concerns raised in these studies is that of teachers' beliefs about the nature of mathematics and mathematics education. This importance can be appreciated by examining the role of these beliefs in the context of some of the research that has been reported here.

Peterson suggests that "Teachers might modify the traditional mathematics curriculum to build systematically on children's informal knowledge." (p. 20, this vol.), a process referred to as "Cognitively-Guided Instruction." One of the principles underlying such an approach is identified as the need for teachers to have "a thorough knowledge of the mathematics content domain" in order to be able effectively to assess students' learning within that domain (Peterson, p. 43). Such a situation provides an example of how teachers' beliefs, both with respect to the nature of mathematics and to how children learn, can be strongly challenged. The "traditional" mathematics curriculum which many teachers are required to teach reflects a view of the discipline that stresses formal content. A formalist view falls within what Wynne (1964) identifies as "formal-discipline" theory where mathematics would emphasise "rules of operation and formal examples" rather than practical problems (p. 11). Cognitively guided instruction, on the other hand, sets out to improve higher-order thinking skills through problem solving. To adopt such an approach clearly may require some teachers to re-examine their views about the nature of the mathematics they teach and the way they teach it. It might be difficult, for example, to learn to value the idiosyncratic nature of children's informal knowledge within the context of the mathematics curriculum and to appreciate that this can lead to more meaningful learning. Thus teachers' long-held beliefs could, in such a situation, be obstructive in implementing potentially desirable changes in a curriculum. Brown and Cooney (1986) point out that "A belief must have some sense of logical or psychological consistency for an individual" (p. 9). The teacher whose beliefs about mathematics are being challenged at the logical level (in terms of what to teach) and at the psychological level (in terms of how to teach) could be doubly confused, yet this is a situation in which teachers are often placed in the context of research and development in mathematics education.

This is a single example of the importance of teachers' beliefs and how they may act in a counterproductive way when curricular change and development are proposed. The opposite, of course, may also be the case. It is possible that by encouraging teachers explicitly to identify their beliefs they may come to appreciate the effect these have on their performance as teachers of mathematics and, if necessary, call them into question. There is often a tendency for those of us in a position to do so not to undertake this task, possibly because we consider it too difficult and doubt the efficacy of carrying out such an exercise. However, Scheffler (1973) offers encouragement in this respect when he says:

> We judge belief in question by its general impact on all other beliefs in which we have some confidence. No matter how confident we are of a particular belief,

we may decide to give it up if it conflicts with enough other beliefs in which we have a higher degree of confidence. (p. 119)

It becomes very important, therefore, to offer teachers a convincing rationale for replacing old beliefs with new in order to generate such confidence. This need is particularly compelling if we wish teachers to adopt a view of themselves and their students as "makers of knowledge" and to understand how this can be so. This necessarily involves the perception of mathematics in more open-ended terms and would necessitate considerable rethinking on the part of many (Nickson, 1986). Alternative theoretical considerations do exist to support such a view (such as those referred to by Bauersfeld), but mathematics teachers should not be expected to search these out and to make such fundamental shifts on their own. Teachers and researchers can be collaborators as Hoyles suggests, but it is important that within that collaboration teachers should be given the support necessary both to justify to themselves any such basic changes and to project the possible implications for what mathematics they teach and how they they teach it.

Finally where beliefs about mathematics and our students are concerned, it is difficult to dispute Mellin-Olsen's (1987) contention that mathematics education can be as important as language education in helping young people to order and make sense of their world and that, as a subject, mathematics is "a structure of thinking-tools appropriate for understanding, building or changing a society" (p. 17). If we really believe this about the subject, we must ask ourselves whether the mathematics curriculum in schools projects such a view. This adds emphasis to the need to integrate our beliefs about the students we teach, the mathematics we teach them, and the strategies we select to do so, as Good and Biddle suggest.

Strategies for the Improvement of Classroom Practice

Leinhardt and Berliner have identified interesting features of how teachers behave in the mathematics classroom. The contrast in approach and conclusions of their respective studies was reflected in much of the discussion that followed the presentation of each of these papers. The notion of the "expert," it seems, is a contentious one and means different things to different people. The suggestion that a person's view of what constitutes an expert teacher is determined by what that individual values seems a valid one. However, a further suggestion that possibly the same teacher will "look" different when teaching different content raises some interesting issues. The managerial and organisational demands involved in teaching different mathematical topics is clearly different, as these two studies have shown. The strategies selected for effective teaching, however, cannot be based only on teachers' perceptions of content as determined by the level of student achievement. The nature of that content clearly must also affect what the teacher does so that the "flow" alters according to topic boundaries, as Leinhardt suggests, rather than according to a predetermined time schedule. Content, therefore, is an important variable that cannot be ignored.

Another focal point for research into strategies for improving classroom practice may be to study teachers working collaboratively in a classroom setting. Participation in collaborative experience of this kind has been identified as the most effective strategy for bringing about change (Assotiation of Teachers of Mathematics, 1987). Clearly, this is built on the notion that to be an effective teacher of mathematics is not a rigid state but requires continuous evaluation and flexibility, and the adaptation of particular content and methods to particular students within a particular setting. This view of the effective mathematics teacher may appear to be counter to the notion of the expert teacher; on the other hand, it would be one way to build on such expertise, if it is identified, and to maximise it as much as possible.

The Mathematics Curriculum, Teacher Autonomy and Professionalism

The notions of autonomy and professionalism merge perceptibly when discussing the content of classroom mathematics lessons. The autonomy of teachers, as selectors of what mathematics is to be taught and to whom, must be potentially the most powerful aspect of their teaching role. Porter has identified the idiosyncratic approach to the selection of content adopted by teachers in his study and suggests that, on the whole, teachers' reasons for selecting particular content are largely unexamined. At the same time, as Romberg points out, an important category of the knowledge teachers have that characterises them as professionals is knowledge of mathematical content and how it relates both to different areas within mathematics itself and to disciplines outside it. Thus, in order to be autonomous in the professional context, the mathematics teacher should be in a position to make informed decisions with respect to what mathematics is taught for its own sake but also what is relevant and necessary, for example, for the study of science. If this kind of decision making is not taking place, it is vitally important to ask why it is not. It is clear that the notion of teacher autonomy within the context of the mathematics classroom can be taken too far and that many teachers would welcome intervention and guidance of an appropriate sort. Rather than resenting it, the teachers in Porter's study reacted in a positive way to the interest taken in what they taught in mathematics lessons. Schofield's study provides evidence of a similar sort, where clearly at least some of the teachers observed would have welcomed constructive intervention in the early stages of using computers in the teaching of mathematics.

The situation is one that demands guidance of a professional kind that can be catered for by offering leadership within the school. This is a move which has been made in the U.K., for example, where the role of mathematics co-ordinator has been established at primary level. This has been done in the belief that "the effectiveness of the mathematics teaching in a Primary school can be considerably enhanced if one teacher is given responsibility for the planning, co-ordination and oversight of work in mathematics throughout the school" (Committee of Inquiry into the Teaching of Mathematics in Schools, 1982, p. 104). At secondary level such guidance is given by the head of department whose role again is seen in terms of "sustained leadership" (p. 154). It has been argued that by identifying a leader in this context the notion of teachers of mathematics as a specific group is emphasised and, hence, the commonalities of their professional concern become shared (Bishop & Nickson, 1983). Leadership of this kind, when appropriately offered, also helps teachers of mathematics to overcome their individual anxieties about the subject in a nonthreatening way.

Imbalance in the selection of mathematical content within the curriculum has been noted elsewhere (e.g., Ashton, Kneen, Davies, & Holley, 1975; Department of Education and Science, 1978; Freeman & Kuhs, 1980). One solution to the problem of avoiding such an imbalance may lie in altering the overall perspective of mathematical content. One such view is the "process" view referred to by Bell, Costello, & Kuchemann (1983), and currently being investigated and developed by the Oxford Certificate of Educational Achievement (OCEA) project in mathematics. This stresses the centrality of processes in the mathematics curriculum and uses these as focal points around which to group the selection of content. The aim is to ensure that fundamental aspects of mathematics, such as generalisation, problem solving, representation, proof, and abstraction, are developed and emphasised in a variety of contexts. In stressing process as opposed to content the mathematics curriculum may become more open-ended in terms of content, but it is believed that because of the nature of the processes, there is more likely to be balance within *kinds* of content selected. The idea of process also suggests links with activity and helps to present mathematics as an area of the curriculum in which students are actively engaged in *doing* mathematics as opposed to passively receiving something that is labelled "mathematics." A key feature of an emphasis on processes is that it is seen as a way to stengthen the development of problem-solving skills.

The result of the sort of approach being described here appears not unlike what Good, Grouws, and Ebmeier (1983) have called "active teaching." A concentration on process rather than on content tends to shift the teacher's attention from materials to strategies, a move implicitly advocated by Good and Biddle in their study. The "give and take" between teacher and student which they describe might be realized more easily and more often if a processes approach to mathematical content were made more accessible to teachers. The importance of "opening up" the mathematics curriculum in this way cannot be overemphasised, but it will necessarily involve a change in perceptions and beliefs about mathematics on the part of many teachers. What this suggests is the principle of *sharing* within the mathematics curriculum, which is a valuable notion when considered at a variety of levels. It links strongly with the provision of leadership which ideally can offer sharing at one level. But teachers also need to have the confidence to share their mathematical knowledge (including their anxieties about it) with each other, something that the collaborative working situations referred to earlier will allow. This in turn could help them to share in a more meaningful way in the "give and take" situations with their students, including sharing their anxieties about mathematics which they as teachers have themselves some-times felt. Perhaps somewhat ironically, the professionalism of mathematics teachers sometimes may involve having the confidence to say "I don't know" in front of students and colleagues alike, and to engage together in finding an answer.

The Role of Theory in Mathematics Education

Theory has been central to the concern of studies presented here (and in much of the conference debate) since theories are what research is about. Bauersfeld reminds us of the complexity of theoretical enterprise generally when he notes that every theory entails a particular perspective and that in adopting this perspective, the theorist necessarily makes a choice to select some aspects of the situation being studied and to disregard others. Thus no theory can accurately reflect the "reality" it is being used to model and to probe. To be reminded of the relativism involved in being objective is a salutary experience. The question persists, however, as to how theory can contribute to our understanding of what happens in the mathematics class-room and how it can help to achieve more effective mathematics education.

A first step towards helping to make this link more explicitly and achieving greater effectiveness has to do once again with sharing, but in this case the sharing is incumbent upon researchers. We need to learn to share our theories with teachers to a far greater extent than we have tended in the past. Clearly, particular kinds of theory may not lend themselves to this, and the research paradigm may preclude it. However, getting theory into classrooms must be a key item on any research agenda for effective math-ematics teaching and, if this is accepted, it follows that it is necessary to adopt or develop research paradigms that allow this to happen.

Sharing ideally is a two-way process and as such, it is important that researchers tell mathematics teachers about their theories and listen to what teachers have to say about these ideas. It is equally important, however, that we listen to *their* ideas, to give them credence and to be seen to value them. Porter's reference to teachers' reactions to the interest of others in what mathematics they taught provides an example of the potential of this kind of sharing when some of those teachers stated their intention to monitor future lessons for their own benefit. This is a good example of how a researcher's theory can come close to the reality of changing classroom practice and lead teachers to reflect upon what they are doing and, hopefully, why. Similarly, teacher-generated theories cannot be ignored for they provide the practitioner's perspective and, in this case, it is the researcher who should be the listener. We cannot afford to disregard the professional instinct of the teacher in helping to identify what can help or hinder effective mathematics education. Good and Biddle point out that better and more persuasive theories include explanations for events, and that the fundamental task of classroom research is to generate and test these theories. The provision of a collaborative climate in which teachers can share ideas, with colleagues from within school and with researchers from outside it, can only increase the potential number of plausible theories for exploration and testing. We would also do well to remember that teacher-generated theories may have the advantage of being more readily accessible to other teachers. For example, the efforts of a teacher who undertakes the systematic involvement of parents with their child's mathematics curriculum and studies the effects of this initiative is unlikely to go unnoticed by her or his colleagues. Perhaps as researchers we are too ready to assume an implausibility about teacher-generated theories that belies our own status as theorists.

Conclusion

By focussing on global issues raised throughout this volume it has been possible to identify more particular ones that could form part of a research agenda for effective mathematics teaching. These have included:

1. the need for within-school leadership for teachers of mathematics;

2. the need for joint development with teachers of a rationale for the curricular changes they are asked to undertake;

3. the identification of strategies to introduce more collaborative teaching and learning situations in mathematics education;

4. the exploration of ways of taking theory into the classroom, in particular through the pursuit of teacher-generated theories.

Issues that have not been touched upon specifically but are implicit in some of the identified concerns are: the accountability of mathematics teachers, the greater involvement of parents in their children's mathematics education, and the elementary/high school interface.

There has been an emphasis throughout this paper on the notion of sharing and this also characterizes the above conclusions relating to potentially valuable areas of research. These conclusions suggest a shift from a concern with the individual in the act of teaching or learning to a concern with individuals and the social interaction within the mathematics classroom. The social context that results from this interaction is susceptible to many factors and influences, not least important of which are the beliefs about mathematics and mathematics education that we bring to the classroom. If we wish to ensure that we teach the subject effectively, we must be aware of how such factors can aid or obstruct such teaching. It may be that "sharing" at a variety of levels in the context of the mathematics classroom may lead to more effective teaching and learning, but we need to find out how to bring this about.

REFERENCES

Association of Teachers of Mathematics. (1987). *Co-ordinating mathematics in primary and middle schools*. Derby, UK: ATM King's Chambers, Queen St.

Ashton, P., Kneen, P., Davies, F., & Holley, B. (1975). *The aims of primary education: A study of teachers' opinions*. London: Macmillan Education Ltd.

Bell, A. W., Costello, J., & Kuchemann, D. (1983). *Research in learning and teaching: A review of research in mathematical education* (Part A). Windsor, UK: NFER-Nelson.

Berger, P., & Luckman, T. (1976). *The social construction of reality*. Harmondsworth, UK: Penguin Books Ltd.

Bishop, A. J., & Nickson, M. T. (1983). *The social context of mathematics education: A review of research in mathematical education* (Part B). Windsor, UK: NFER-Nelson.

Brown, S. I., & Cooney, T. J. (1986, November). *Stalking the dualism between theory and practice*. Paper presented at the conference on Systematic Cooperation Between Theory and Practice in Mathematics Education, Lochem, The Netherlands.

Committee of Inquiry into the Teaching of Mathematics in Schools, (1982). *Mathematics Counts* (The Cockcroft Report). London: Her Majesty's Stationery Office.

Department of Education and Science (1978). *Primary education in England: A survey by HM Inspectors*. London: Her Majesty's Stationery Office.

Freeman, D. J., & Kuhs, T. M. (1980, April). *The fourth grade mathematics curriculum as inferred from textbooks and tests*. Paper presented at the annual meeting of the American Educational Research Association, Boston.

Good, T., Grouws, D., and Ebmeier, M. (1983). *Active Mathematics Teaching*. New York: Longman.

Mellin-Olsen, S. (1987). *The politics of mathematics education*. Dordecht, Holland: D. Reidel.

Nickson, M. T. (1986, December). *Towards a multicultural mathematics curriculum*. Paper presented at the conference on Social Perspectives of Mathematics Education, University of London Institute of Education.

Scheffler, I. (1973). *Reason and teaching*. London: Routledge & Kegan Paul.

Wynne, J. P. (1964). *Theories of education*. New York: Harper & Row.

An Agenda for Research on Teaching Mathematics

Thomas J. Cooney
University of Georgia
Douglas A. Grouws
University of Missouri
Doug Jones
University of Georgia

THE EVOLUTION OF AGENDAS

After participating in an agenda-setting conference and reading the papers in this monograph, it is difficult not to reflect on the directions in which the profession has moved and is moving with respect to research on teaching mathematics. We have, it seems, come a long way, at least in terms of the complexity of the questions and issues being addressed. At a previous agenda-setting conference, held in the spring of 1975, mathematics educators grappled with the issues of what a feasible agenda for research on teaching mathematics might be. The framework of that conference consisted primarily of the work of Kenneth Henderson and B. O. Smith and their work on "moves" for teaching mathematical concepts, generalizations, and skills. The conference resulted in the identification of three different basic thrusts for research: (a) analyses of classroom discourse; (b) analyses of effective sequences of moves for teaching concepts, generalizations, and skills; and (c) development of strategies for using the sequences of moves in teacher education programs. Considerable research in these areas was initiated following the conference. This line of research gave attention to "teachable objects," which were specific types of mathematical content such as concepts, principles, and skills (see Cooney, 1980, for a review of this work). These studies were important because previous research on teaching behavior had not given sufficient attention to the nature of the mathematics being taught. A criticism of this research was its tendency to compartmentalize mathematics, thus obscuring a more holistic perspective about mathematics. There were, of course, other kinds of research on teaching mathematics taking place during this period. Much of this research involved process-product studies in which teaching behavior in mathematics classrooms was observed and linked with student learning outcomes, typically student achievement (see Brophy and Good, 1986, and Romberg and Carpenter, 1986 for a review of this research). Initially, these studies often took little account of the context in which teaching occurred. For example, differences in curriculum, instructional pace, SES, and types of mathematics lessons were insufficiently considered both in conducting the research and in interpreting the results. More recent studies have alleviated these shortcomings and have added an experimental loop in the research design in order to begin to establish causal effects. Through such efforts a number of useful constructs, such as the development portion of a lesson (Good, Grouws, & Ebmeier, 1983), have emerged. Yet this line of research has limitations, too, because it relies on existing practice in the schools and thus can only bring instruction up to the best of currently available practice, rather than make instruction radically different. The variety of backgrounds of the individuals participating in this conference suggests that communication among educational psychologists, cognitive scientists, and mathematics educators has increased substantially in the last few years. This increase in communication holds the potential for the development of insightful ways to move the understanding of good mathematics teaching forward.

Missing Links

When we consider the state of our knowledge about ways of thinking about research on teaching mathematics at the time of Begle's (1979) *Critical Variables*, we realize how meager our concepts for thinking about research on teaching really were. Perhaps a contributing factor to this state of affairs is the fact that there was not a rich background of classroom observational studies on teaching mathematics, as noted by Good and Biddle in this monograph. Observations are necessary for developing perspectives, for identifying potent variables, and for developing a language for talking about children's and teachers' discourse in the classroom and the thought processes that underlie their discourse. Without such observations we are left with research that has minimal potential for developing models for conducting research in mathematics education.

The work of Porter and his colleagues presented in this monograph is important for the very reason that it attempts to document relationships that are often assumed in observational studies. The determination of what affects teachers' selection and interpretation of content is immensely important if the profession is to consider various means of reform. The issue can be addressed both at the level of whether or not geometry is taught and at the more detailed level, such as why fractions are interpreted as regions by one teacher, as decimals by another, and as points on a number line by still another. What factors contribute to different interpretations? One factor is certainly teachers' beliefs. To what extent are interpretations or emphases driven by testing programs?

One of the authors of this paper conducted an in-service program with a group of teachers who were concerned with raising local test scores on a statewide test. The teachers had little information about the content of the test, but they were convinced that computational aspects were very important, especially with respect to fractions. They were quite surprised that released items revealed an emphasis on conceptual interpretations of fractions far more than on computational aspects. While mathematics educators are quick to argue that testing drives the curriculum and the teaching of the curriculum, the fact is that we have little documentation of just how and in what contexts that occurs, or of what factors should be addressed in order to reverse the situation.

Although development of higher-order thinking skills has been the centerpiece of NCTM's *Agenda for Action*, we have little information about what constitutes higher-order thinking skills. We also have little information on how to help teachers teach for higher-order outcomes. In some sense we appear to assume that once we learn how students think about mathematics we will have somehow solved the analogous problem of how teachers think about teaching mathematics and how instruction should be structured. Studies by McGalliard (1983) and Kesler (1985) suggest that teachers communicate a rather dualistic, cut-and-dried conception of mathematics, one that clearly does not foster students' higher-order thinking skills. Thompson (1982, 1984) observed that the conceptions of mathematics and of mathematics teaching held by teachers are at least a partial determiner of how mathematics is taught.

Clearly, questions related to improving the teaching of mathematics are not resolved only by studying how students learn mathematics. Peterson's paper is particularly interesting in that her research, conducted with colleagues at the University of Wisconsin, has a teacher education component that addresses the issue of how knowledge of students' thinking can be used effectively in a teacher education program. Such linkages are immensely important to our profession.

The Importance of International Perspectives

Stigler and Perry's paper in this monograph enables us to view ourselves from the vantage point of different cultures. When we consider the attitudes of parents and teachers from other cultures, we can begin to see why the culture of schooling in the United States may impede the development of the types of professional teachers described by Romberg. Consider, for example, the implication of the contrasting perspectives that students in the United States learn mathematics because of ability, whereas Japanese students learn because of effort. Consider also the notion of what it means to be an "expert" mathematics teacher. Students of expert teachers seem to be "highly proficient in performing the procedures and computations associated with a particular piece of mathematics and in learning the next level of mathematical material presented to them, and moderately proficient in explaining, generalizing, and coping with a variety of extension and higher-order tasks" (Leinhardt, this volume, p. 47). It would be interesting to learn how mathematics educators from other cultures define "expert teachers." For example, it may be that Japanese teachers tend to emphasize process over product and to demonstrate a certain flexibility when solving mathematical problems. According to Stigler and Perry, Japanese classrooms tend to emphasize more reflective thinking than those in the United States. The point is that our definition of what is expert reflects the culture in which we live. Consideration of expertise as defined in a different culture might lead us to consider different conceptualizations.

And so, we see that research on teaching mathematics has progressed, at least in the complexity of the issues addressed in developing an agenda for future research. Although theoretical perspectives for viewing the problems are still eclectic and basically empirically driven—as perhaps they should be—we are at least in a position to realize a better map of the terrain for conducting research than we were a decade or so ago. While some might argue that this is rather slow progress for a decade of work, we do not share this view. Often the identification of questions and the creation of methods for addressing those questions represent significant progress. Consider, for example, the resulting significance for mathematics and for the development of non-Euclidean geometries when the question about proving the parallel postulate shifted from "How" to "Whether."

CURRENT AGENDAS

It is important to remember that this monograph reports the results of a *conference* and is not just a collection of papers. A vital and important aspect of the conference was the human interaction. As a result, we have a rich and generative base upon which to begin to build an agenda for research on effective mathematics teaching. It should be emphasized that we are not proposing a set and static agenda. As John Dossey, president of NCTM, pointed out at the closing agenda-setting session of the conference, "The *process* [of agenda setting] *begins* here, it doesn't end here." Indeed, the participants were operating from that perspective throughout the conference.

In order to capture both the human element and the agenda-setting tone of the conference, we offer the following comments from some of the participants.[*]

Content

As noted earlier, research on teaching mathematics has not always focused on the nature of mathematics. However, at this conference, it emerged as a major concern:

Research on teaching mathematics needs to be concerned with what is important mathematics to teach.

There was also concern with the integration of mathematics, teaching, and the curriculum: We need to look at what mathematics the teacher knows that is *relevant* to the content we have decided is important, not just at knowledge of specific content.

We need to incorporate context into the research and into the teaching. What effect does the subject of mathematics have on the classroom?

Too many students experience mathematics in a rather disjointed way. That is, their mathematical studies are composed primarily of *lessons*. Such students might not find the joy of organizing mathematics into a coherent whole. This prompted one participant to suggest that:

> We need to look at it from less a *lesson* perspective and more a *general-overview-of-mathematics* perspective. We need teachers to have a more holistic understanding of mathematics.

Beliefs

Recent research indicates the importance of considering the personal, value-laden worlds of mathematics teachers' and students' beliefs—about mathematics, about teaching, and about teaching mathematics. Participants suggested that use of this information could help teachers at all levels to reflect on their teaching and could help students to become more aware of their own learning: We need to study teachers' cognitions. What do they think about when they are teaching?

> We need to study students' cognitions. What are they doing and why? What are the students' perceptions of the *coherence* of the class and of individual lessons.

The direction called for at the conference was the intentional use of findings from this important line of research: We need to help teachers to identify their conceptions of mathematics. How do these conceptions affect teaching?

> We must look at attitudes, beliefs, and knowledge. We must look for ways to try to work this into teacher education.

Experts and Novices

A considerable amount of discussion focused on research related to expert/novice teachers. The view was expressed that it is important to have a knowledge base about what differentiates experts and novices and how the novice can become more "expert-like" through various teacher education programs. One of the problems addressed was the means by which we can help the nonexperts develop expertise. As one participant put it:

> What happens to promote the transition from novice to expert? It is important for our field to study this.

The question of how to designate experts and novices was considered. In addition, one participant challenged the group to consider the question, "Expert with respect to what?"

The question was also raised as to what is being compared when one defines *expert* and *novice*. The following question was posed.

> Can you really compare experts and novices? There is a lot that confounds the issues and makes them difficult to unravel.

One of the confounding factors is experience. Although a teacher with 10 years of experience may not be an expert, it seems unreasonable to label him or her a novice. However, it is not unreasonable to believe that a teacher could become an expert with very few years of experience. There was the question, as well, of how the beliefs expert teachers hold about mathematics and the teaching of mathematics might differ from those of the novice, and what implications those differences might have for teacher education.

Teacher Education

There has been much research on how children learn mathematics. However, there are at least three other areas of equal importance in mathematics teacher education:

> There are four levels we need to pay attention to: teaching mathematics, learning mathematics, learning to teach mathematics, and learning to teach teachers of mathematics.

All four of these are important for helping us to conceptualize the mathematics classroom. Whether we think of active mathematics teaching, of reflective practitioners, of involving students as apprentices in "mathematics workshops," or of something else, our conceptions are enhanced by constructing models of the teaching-learning process.

> Model building is important. It can help to guide our research and may establish a "feedback" system. The expectations, attitudes, and conceptions of teachers and of students feeds in, and helps us to create better models. We can then take the models back out to the field.

The integration of theory and practice implicit in this method for constructing models is a critical part of any agenda for research on effective mathematics teaching.

While the theme of the conference was effective mathematics teaching, discussion was never very far from teacher education:

> What is an effective mathematics teacher education program?
>
> Do mathematics educators integrate research on teaching into their own teaching, particularly to teacher training?
>
> We need to consider the teacher education perspective. [For example,] what is the transfer of classroom pedagogical knowledge gained from preservice teacher education programs to work in the field?

Perhaps one way to influence this transfer would make use of a more clinical setting in which our students could practice teaching:

> We need more laboratory settings—instructional laboratories. We need the ability to dissect lessons.

The implication is that the opportunity to look back in detail at what transpired during a mathematics class might enable student teachers to see that class from a different perspective. Having the possibility for structured and purposeful reflection would undoubtedly enhance our teacher education programs. Some professions, such as law and medicine, have built up a literature that deals with applications of the profession to specific cases. It was suggested that mathematics teacher education programs could benefit from having a similar literature:

> There is a need for case knowledge, of both good and bad lessons. This would include transcripts, student performance on various tasks, tests, videos, analysis by professional mathematics educators, et cetera. All of this needs to be pretty specific; for example, equivalence of fractions. We need this kind of case knowledge for all of the common themes.

Making contact with such a case literature could add not only to the practical knowledge that preservice teachers are developing, but also to the theoretical bases for teaching. There was discussion about the lack of research on effective mathematics teaching at the secondary level:

> We must stress the need for focusing on the secondary level and beyond.

In recognizing that mathematics is not just a placeholder in our mathematics teaching, we must come to grips with the different needs of elementary, middle, secondary, and postsecondary teachers. There was concern that practicing teachers need the ongoing support of our teacher education programs:

> We need a model of staff development that is theoretically based. We must have training, then follow-up observation.

Collaboration with Practicing Teachers

There were important discussions dealing with the role of teachers in the research process:

> We need to break down the isolation of teachers.

One way to help alleviate that isolation is to encourage greater involvement in the profession:

> It is important for teachers to be able to discuss things professionally.

A common complaint about research is that it is of little value to teachers in the field. The language is often somewhat technical, the methodology and the statistical or qualitative analyses are difficult for nonresearchers to understand, and, of most importance, the situations described and the suggestions made are thought to have little contact with the reality of the classroom. This perception tends to isolate the teacher from research that might potentially be of help. In short, the research process needs to be more accessible to teachers. One way to do this is to involve teachers not only in the research process, but also in theory building, in the interpretation of results and observations, and in formulating the implications of the research.

Paradigms of Research

A primary goal of this conference was to begin to set an agenda for further research on effective mathematics teaching. Inherent in that task is the formulation of conceptions of research that require examination.

Participants felt there is a need to consider an eclectic approach to research on teaching, both in our theoretical underpinnings:

> Analyzing classrooms requires using many theoretical frameworks. . . . Classrooms are so rich and complex that we must take an eclectic approach to our investigations. This is very hard, but necessary. When we focus on one thing, others fade away.

and in our methodology:

> We should have cross-disciplinary research teams. This can empower the lens.

Finally, there was concern with the basic questions, "What are the goals we have set for education?" and, "Why do we engage in research?":

> We've acted as if our upper bound responsibility is to define terms. What about autonomy and personhood implications in our own research?
> We need to be concerned with the question of why we do the research we do. The bottom line is concern for children and for learning.

CLOSING PERSPECTIVES

Progress in educational research is related to how well the research permits the formation of constructs and theories that are useful in generating hypotheses, in informing classroom practice, and in influencing educational policy. In an effort to promote such progress, many different perspectives about research were expressed during the conference. Some researchers saw progress as being achieved through investigations in which research attempts to determine and describe reality, thereby identifying generalizations that contribute to progress in educational research. Other researchers tended toward an epistemology expressed by Bauersfeld that emphasizes a paradigm shift away from seeing science as a process of determining reality and toward a view of science as a process of considering how a researcher's perspective "fits" with the perspective of those being studied.

In addition to what the researcher considers to be appropriate research, it seems reasonable to assume that the researcher's conceptions of mathematics and of effective methods of teaching mathematics are also influencing factors in shaping research agendas. Although Treffers (1987) was referring more to instruction than to research, his words seem appropriate for developing research agendas as well.

> It appears that starting points and goals are no trivial formulations—which is of course no surprise—but that they in fact yield consequences for the contents of instruction and even for theory-building. (p. 216)

Conference participants consisted of mathematics educators, educational psychologists, cognitive scientists, social psychologists, and mathematicians. It follows that perspectives about epistemological considerations and philosophical perspectives about mathematics varied and influenced individual research agendas in varying ways and degrees. Regardless of individual orientations, the participants shared the common goal of developing constructs that can stimulate teachers' thinking and improve the process of schooling.

It is our impression that the conference presentations, reactions, and discussions provided each of the participants with a basis for reflecting on issues related to the selection and use of paradigms in an effort to better define individual research agendas. We hope that this monograph provides its readers with a similar experience as they contemplate the development and furthering of their own research agendas for promoting progress in educational research.

REFERENCES

Begle, E. G. (1979). *Critical variables in mathematics education*. Washington, DC: Mathematical Association of America.

Brophy, J. E. & Good, T. L. (1986). Teacher behavior and student achievement. In M. C. Wittrock (Ed.), *Handbook of research on teaching* (3rd ed.)(pp. 328-375). New York: MacMillan.

Cooney, T. J. (1980). Research on teaching and teacher education. In R. Shumway (Ed.), *Research in mathematics education*. Reston, VA: National Council of Teachers of Mathematics.

Good, T. L., Grouws, D. A., & Ebmeier, H. (1983). *Active mathematics teaching*. New York: Longman.

Kesler, R. (1985). *Teachers' instructional behavior related to their conceptions of teaching and mathematics and their level of dogmatism: Four case studies*. Unpublished doctoral dissertation, University of Georgia, Athens, Georgia.

McGalliard, W. A. (1983). Selected factors in the conceptual systems of geometry teachers: Four case studies. (Doctoral dissertation, University of Georgia, Athens, Georgia 1983.) *Dissertation Abstracts International 44A*: 1364.

Romberg, T. A. and Carpenter, T. P. (1986). Research on teaching and learning mathematics: Two disciplines of scientific inquiry. In M. C. Wittrock (Ed.), *Handbook of research on teaching*(3rd ed.) (pp. 850-873). New York: MacMillan.

Thompson, A. G. (1982). *Teachers' conceptions of mathematics and mathematics teaching: Three case studies*. Unpublished doctoral dissertation, University of Georgia, Athens, Georgia.

Thompson, A. G. (1984). The relationship of teachers' conceptions of mathematics and mathematics teaching to instructional practice. *Educational Studies in Mathematics*, *15*, 105-127.

Treffers, A. (1987). *Three dimensions*. Dordrecht: D. Reidel.

Working Group on
Teaching and Assessing Problem Solving

San Diego, California
January 9–12, 1987

Joan Akers
San Diego County
  Office of Education

John Bransford
Vanderbilt University

George W. Bright
University of Houston

Ann L. Brown
University of Illinois

Thomas P. Carpenter
University of Wisconsin-
  Madison

**Randall I. Charles
Illinois State University

Clyde Corcoran
Whittier High School
  District

John Donald
San Diego State University

*James G. Greeno
University of California,
  Berkeley

*Jeremy Kilpatrick
University of Georgia

Gerald Kulm
AAAS

Jean Lave
University of California,
  Irvine

Frank K. Lester
Indiana University

Sandra P. Marshall
San Diego State University

*Douglas B. McLeod
Washington State University

Nel Noddings
Stanford University

Nobuhiko Nohda
University of Tsukuba

Tej N. Pandey
California Assessment
  Program

Lauren B. Resnick
University of Pittsburgh

*Thomas A. Romberg
University of Wisconsin

Alan H. Schoenfeld
University of California,
  Berkeley

Richard J. Shavelson
University of California, Los
  Angeles

**Edward A. Silver
San Diego State University

*Judith T. Sowder
San Diego State University

Larry Sowder
San Diego State University

George M. A. Stanic
University of Georgia

*James W. Stigler
University of Chicago

Alba G. Thompson
Illinois State University

James W. Wilson
University of Georgia

Working Group on
Effective Mathematics Teaching

Columbia, Missouri
March 11–14, 1987

Heinrich Bauersfeld
Universitat Bielefeld

Jacques C. Bergeron
Université de Montréal

David C. Berliner
University of Arizona

Bruce J. Biddle
University of Missouri

Catherine A. Brown
Virginia Polytechnic Institute

Stephen Brown
State University of New
  York-Buffalo

William S. Bush
University of Kentucky

**Thomas J. Cooney
University of Georgia

John A. Dossey
Illinois State University

Elizabeth Fennema
University of Wisconsin-
  Madison

Sherry Gerleman
Eastern Washington State
  University

Thomas L. Good
University of Missouri

**Douglas A. Grouws
University of Missouri

Celia Hoyles
University of London

Martin L. Johnson
University of Maryland

Mary Koehler
University of Kansas

Perry E. Lanier
Michigan State University

Gaea Leinhardt
University of Pittsburgh

Richard Lodholz
Parkway Public Schools

Marilyn Nickson
Essex Institute of Higher
  Education

John Owens
University of Alabama

Penelope L. Peterson
University of Wisconsin

Andrew C. Porter
Michigan State University

Edward Rathmell
University of Northern Iowa

Laurie Hart Reyes
University of Georgia

*Thomas A. Romberg
University of Wisconsin

Janet W. Schofield
University of Pittsburgh

Robert Slavin
Johns Hopkins University

*Judith T. Sowder
San Diego State University

*James W. Stigler
University of Chicago

Working Group on
The Learning and Teaching of Algebra

Athens, Georgia
March 25–28, 1987

John E. Bernard
West Georgia College

Lesley R. Booth
James Cook University-
Queensland

Diane J. Briars
Pittsburgh Board of
Education

Seth Chaiklin
Bank Street College

Robert B. Davis
University of Illinois

James T. Fey
University of Maryland

Eugenio Filloy Yague
Centro de Investigacion y
Estudios Avanzados del
I.P.N.

Larry L. Hatfield
University of Georgia

Nicolas Herscovics
Concordia University-
Montreal

Robert Jensen
Emory University

Mary Grace Kantowski
University of Florida

James J. Kaput
Southeastern Massachusetts
University

**Carolyn Kieran
Université du Québec a
Montréal

David Kirshner
University of British
Columbia

Jill H. Larkin
Carnegie Mellon University

Joan R. Leitzel
Ohio State University

Matthew Lewis
Carnegie Mellon University

Tatsuro Miwa
University of Tsukuba

Sidney Rachlin
University of Hawaii

Sharon L. Senk
Syracuse University

*George Springer
Indiana State University

*Judith T. Sowder
San Diego State University

*Jane O. Swafford
Northern Michigan
University

David Tall
University of Warwick

Patrick W. Thompson
Illinois State University

John A. Thorpe
National Science Foundatic

**Sigrid Wagner
University of Georgia

David Wheeler
Concordia University-
Montreal

Patricia S. Wilson
University of Georgia

Working Group on
Middle School Number Concepts

DeKalb, Illinois
May 12–15, 1987

**Merlyn J. Behr
Northern Illinois University

Alan Bell
Shell Centre, Nottingham

Robbie Case
Ontario Institute for Studies
in Education

Karen C. Fuson
Northwestern University

Brian Greer
Queens University, Belfast

Kathleen M. Hart
Kings College, London

**James Hiebert
University of Delaware

Thomas E. Kieren
University of Alberta

Magdalene Lampert
Michigan State University

Glenda Lappan
Michigan State University

Richard Lesh
WICAT

Jack Lochhead
University of Massachusetts

*Douglas B. McLeod
Washington State University

Pearla Nesher
University of Haifa

Stellan Ohlsson
University of Pittsburgh

Joseph N. Payne
University of Michigan

Thomas R. Post
University of Minnesota

Robert E. Reys
University of Missouri

*Thomas A. Romberg
University of Wisconsin

Judah L. Schwartz
Education Technology
Center

*Judith T. Sowder
San Diego State University

Leslie P. Steffe
University of Georgia

Gérard Vergnaud
Greco Didactique, Paris

Ipke Wachsmuth
Universitat Osnabruck

Diana Wearne
University of Delaware

*Advisory Board members and Project Director
**Conference Co-directors